Telecommunication Services in Sub-Saharan Africa

AF204442

DEVELOPMENT ECONOMICS
AND POLICY

Series edited by Franz Heidhues and Joachim von Braun

Vol. 26

PETER LANG

Frankfurt am Main · Berlin · Bern · Bruxelles · New York · Oxford · Wien

Telecommunication Services in Sub-Saharan Africa

An Analysis of Access and Use in the Southern Volta Region in Ghana

Romeo Bertolini

PETER LANG

Europäischer Verlag der Wissenschaften

Die Deutsche Bibliothek - CIP-Einheitsaufnahme

Bertolini, Romeo:

Telecommunication services in Sub-Saharan Africa : an analysis
of access and use in the southern Volta region in Ghana / Romeo
Bertolini. - Frankfurt am Main ; Berlin ; Bern ; Bruxelles ; New
York ; Oxford ; Wien : Lang, 2002
    (Development economics and policy ; Vol. 26)
    Zugl.: Bonn, Univ., Diss., 2001
    ISBN 3-631-39161-7

Gedruckt mit Genehmigung der
Mathematisch-Naturwissenschaftlichen Fakultät
der Rheinischen Friedrich-Wilhelms-Universität
Bonn.

D 5
ISSN 0948-1338
ISBN 3-631-39161-7
US-ISBN 0-8204-5498-2
© Peter Lang GmbH
Europäischer Verlag der Wissenschaften
Frankfurt am Main 2002
All rights reserved.

www.peterlang.de

Preface

The overall question that is asked in this research by Romeo Bertolini is: Who uses telecommunications in rural areas, what determines this use and how do people benefit from it?

It is often argued that by bridging the information gap through spreading information and communications technologies (ICTs), economic growth will be accelerated, agricultural and industrial productivity as well as the efficiency of public administration will be increased and the competitiveness of developing countries strengthened. However, there are thus far only few empirical analyses. This leaves room for scepticism: Some point out that most infrastructural developments in the ICT field are restricted to the urban centres of developing countries, which fosters the information gap at the sub-national level and further polarises urban and rural areas. Others recognise access to the new technologies as a function of basic socio-economic factors. The inability to access ICT, due to limited education or inappropriate language skills, and the prevalence of inequalities in access will tend to intensify information gaps and thus increase inter-personal and inter-regional inequalities. The probably most challenging issue for development research and policy may be that developing countries have other, more pressing investment priorities. Devoting limited resources to ICTs is, according to some, difficult to justify, especially if there is only inadequate evidence of beneficial returns.

The latter issues are at the centre of interest of a series of research work done at ZEF, the Centre for Development Research. The work presented in this volume needs to be seen in the context of this series and as an attempt to confront the uncertainty around ICT for Development with the situation of households in rural areas of sub-Saharan Africa. Beyond descriptive and univariate analyses of households' use of ICTs as they are provided by public call offices and boxes, it tries to set up a conceptual frame which allows to relate the characteristics of households to the utilisation of telecommunication services.

The results of this research uncover the considerable catchment area of the services and the high penetration of use in rural communities. Additionally, they show that people are willing to take high costs in terms of time and money into account in order to be able to utilise the communication services. It is also made clear that the high demand is by no means limited to the economically wealthy and active, although user-rates are particularly low amongst women and illiterates.

Concerning the benefits, the work underlines the importance of strong social ties as well as their social and economic relevance as one major driving force for using the telephone and benefiting from the related time and money savings. Evidence is furthermore given for the need to recognise information as an important input into the intra- and inter-households' economic activities. Increasing the amount of information enables choices and beneficial decision-making processes.

Through such findings, moves towards a better understanding of the mechanisms induced by ICT use were made. It is hoped that such moves pave the way for a more adjusted and differentiated perception of the implementation of ICTs in low-income countries and the risk of excluding vast majorities from these potentials.

Prof. Dr. Joachim von Braun
Center for Development Research (ZEF)
University of Bonn

Prof. Dr. Franz Heidhues
Center for Tropical Agriculture
University of Hohenheim

Acknowledgements

As described in this work, access to telecommunications is to a large extent determined by the political and macro-economic framework. ZEF, the Center for Development Research, and not least this work were just as well enabled by historic changes, hence, the German unification and the need to compensate for Bonn's loss in centrality functions.

This need opened opportunities, at first for only a few but over time for an ever increasing number of researchers from all over the world. It was and still is a great challenge and opportunity to grow with the Center during these early years of its existence.

I especially thank Professor von Braun for opening up this opportunity (and many more) for me. I would also like to express my appreciation to my colleagues from the research team on ICTs for Development, namely Arjun Bedi, Shyamal Chowdury, Gi-Soon Song, Maximo Torero, and Wensheng Wang. I owe particular gratefulness to my office mate Maria Iskandarani and my colleague Dietrich Müller-Falcke for the constructive and enjoyable collaboration, and to Annette van Edig and Clare and Marc Bromiley for their professional editorial support.

Without my colleagues in Ghana, Patience Asem, Owuraku Sakyi-Dawson and Amos Anyimadu, this research would simply not have been possible. Thank you.

I'd like to express my special regards and best wishes to Professor Ehlers for his helpful support and constructive feedback. On a more private note, his restless involvement in developing Bonn towards an internationally known city of science and development co-operation also formed the basis to meet Maarit. I thank her for her great support and patience and her family for many moments of relaxation. Kiitos!

With special emphasis I finally thank my parents Renate und Franco Bertolini for their trust and unceasing support.

Bonn, November 2001   Romeo Bertolini

Table of Contents

List of Figures, Tables, and Boxes

Figures

Tables

Boxes

Acronyms

ADP	Accelerated Development Program
DCD	District Co-ordinating Director
ECG	Electricity Corporation of Ghana
ENOWID	Enhancing Opportunities for Women in Development
FAO	Food and Agriculture Organization
FCFA	Franc de la Communauté Financière Africaine
GATS	General Agreement on Trade and Services
GDP	Gross Domestic Product
GNP	Gross National Product
GNPC	Ghana National Petroleum Corporation
GPRTU	Ghana Private Road Transport Union
GSM	Global Standard for Mobile Communication
GT	Ghana Telecom
GTZ	Gesellschaft für Technische Zusammenarbeit
ICT	Information and Communication Technologies
IDD	International Direct Dialling
IMF	International Monetary Fund
ISP	Internet Service Provider
IT	Information Technology
ITU	International Telecommunications Union
LAN	Local Area Network
LDCs	Least Developed Countries
NCA	National Communication Authority
OECD	Organisation for Economic Co-operation and Development
PC	Personal Computer
PCO	Public Call Office
PPP	Purchasing Power Parity
PTO	Public Telecommunications Operator
R&D	Research and Development
ROSCAS	Rotating Savings and Credit Associations
SAPRIN	Structural Adjustment Participatory Review International Network
SNO	Second National Operators
SONATEL	Operateur de Télécom du Sénégal
SSA	Sub-Saharan Africa

SUSU	Ghanaian Savings Scheme
TAP	Technology Assessment Project
TCP/IP	Transmission Control Protocol / Internet Protocol
UN	United Nations
UNDP	United Nations Development Program
UN-ECA	United Nations Economic Commission for Africa
WAN	Wide Area Network
WLL	Wireless Local Loop
WTO	World Trade Organisation
WWW	World-Wide-Web

1 Introduction

1.1 Problem Setting

For a long period of time, investments in telecommunication infrastructure in low-income countries were seen as serving the convenience of the rich. Telecommunication services were accordingly regarded as consumer goods that represent and eventually increase the gap between the rich and the poor (LEFF 1984).

Along with the global shift towards information-based economies and technological innovations in the field of microelectronics and telecommunication, this situation is significantly changing. There is a growing awareness of the universal applicability of these cross-cutting technologies. Moreover, the growing demand for ICTs in developing countries has become visible from long waiting lists for telephone connections, the success of cellular systems and the expanding number of Internet users.

An increasing number of low-income countries set up ICT strategies, liberalised their telecommunication markets and opened them to foreign investment. In numerous countries, this led to considerable increases of fixed main lines, falling telecommunication prices, and growing markets for mobile communication and value-added, i.e. Internet-related, services (ITU 1998; WORLD BANK 1998). This also applies to Ghana, the main geographical focus of this work. The West African country is often seen as a best practice example when it comes to information and communication sector reforms.

These reforms are often induced by the insight that reasonably well managed telecommunication entities can generate large financial surpluses in local currency (SAUNDERS ET AL. 1994). In addition to that, telecommunication networks are the neural backbone of more sophisticated ICT applications such as the Internet and are therefore regarded as crucial for economic development:

> "Without appropriate telecommunications, whether it be telephone or e-mail, computers or wind-up radios without batteries, poverty will not be fought and equity and opportunity will not be provided. Without access to modern communications, the difference between rich and poor will increase due to the knowledge gap, and growth will be restrained" (WOLFENSOHN 1999, p. 16-17).

This statement should underline a general tendency in policy makers' current argumentation to assume that ICTs are creating a new economy in which information is the critical resource and basis for competition. While this may be open to discussion, the particular concern of development research should be whether and how the spread of ICTs actually promotes economic development and consequently leads to poverty reduction in developing countries.

In this context, the importance of *information gaps* as a development constraint is stressed and asserts that the provision of new technologies will help to bridge the North-South gap. It is, for instance, argued that by bridging the information gap the spread of ICTs will accelerate growth, increase agricultural and industrial productivity, increase the efficiency of public administration and the effectiveness of economic reforms, strengthen the competitiveness of developing countries, and encourage greater public participation and democracy. Despite these claims, potentials of ICTs are often stated axiomatically but rarely described and analysed in a clear manner.

This leaves room for scepticism. It may, for instance, be pointed out that most infrastructural developments in the ICT field are restricted to the urban centres of many countries in the developing world, and to those parts of the population with higher incomes. Poor people, and the inhabitants of rural areas in particular, even lack the most basic telecommunication service infrastructure (ITU 1998; WORLD BANK 1998). Thus, a growing communication gap not only separates the information rich of the North from the information poor of the South, but eventually fosters trends of polarisation between urban and rural areas on the sub-national level with all its socio-economic consequences (SEIBEL ET AL. 1999).

On the user level, access to the new technologies is largely a function of the existing education, income and wealth distributions. It is argued that both the inability to access, due to limited education or inappropriate language skills, and the prevalence of inequalities of access will tend to intensify information gaps and thus increase inter-personal and inter-regional inequalities. Finally, and perhaps most importantly, developing countries may have other, more pressing investment priorities. Devoting limited resources to ICTs is difficult to justify, especially if there is only inadequate evidence of beneficial returns.

The variety of views expressed above suggests that the role played by information and communication technologies is still vague and that the debate around this role suffers from a lack of convincing differentiation, empirical evidence, and information. While this seems ironic, it is clear that if the new technologies are to command the continued interest of the developing world and justify additional investments, more differentiation and evidence of the returns to the use of these technologies are required.

1.2 Objectives and Research Questions

Against this background, this research focuses on the specific problems of rural areas concerning their integration into information and communication networks and addresses the scarcity of analytical frameworks and empirical evidence in this field. Unlike numerous evaluation studies and policy papers, the presented work does not concentrate on the latest technologies such as GSM networks or Internet applications. Instead, it centres around basic telephony.

The overall aim of this dissertation is to show ways how to empirically assess the way basic services are accessed and used by rural households[1] and to analyse how these households benefit from that use. To achieve this goal it is, on the one hand, necessary to consider the economic and political framework that determines the access to the communication infrastructure. On the other hand, the situation and characteristics of rural households which lead to the use of telecommunication infrastructure will be crucial elements of this work.

This requires the explanation of the impact of general economic and political developments and institutional factors on the supply of telecommunication infrastructure. Due to the overall scarcity of such infrastructure in rural areas of low income countries, on-the-ground-supply does not happen in the form of residential access to telephone services but is rather provided by publicly accessible facilities such as communication centres or phone booths.

The use of the services is determined by a number of variables on the local and the user level which go beyond just having the possibility to use telephones. The empirical part of this work will focus on this user level and will show ways to answer the following central questions:
- Who has access to telecommunication services?
- How is this access utilised on the household level in a specific rural setting in Ghana?
- Through what mechanisms and to what extent do rural households benefit from using telecommunication services?

1.3 Methodological Framework

The questions outlined above will be tackled using different methodological tools.

---

1   Within this work, households are generally considered as both, productive and consumptive units.

Concerning the analysis of the service supply, literature dealing with the increasing importance of ICTs for development and associated regulatory questions will be utilised to describe the expected benefits from sector developments and institutional change. The infrastructural consequences of political and legal issues will be analysed on the bases of secondary data derived from international organisations, service providers, and private consultancy firms. On the local level, a primary supply assessment of publicly available telecommunication facilities by an associated research project was used and extended to better serve the needs of the present research work.

On the demand side, the primary data collected not only assesses the penetration and intensity of telecommunication use but also enables us to relate this use to measures of welfare. To do so, two major analytical steps will be carried out. First, a descriptive and bivariate analysis will provide an overall view of the socio-economic situation of the households as well as of the extent to which households actually use telecommunications and in what manner they are used. Second, multivariate analyses will enable us to estimate the direction and intensity of the impact household characteristics have on the decision to use the available services. Similarly, we will determine the intensity of telephone use. From a conceptual perspective, this will be done following TORERO (2000) and MITCHELL (1995) who modelled the demand of rural households for telephone services by concentrating on access estimation.

The benefits which are expected on the household level mainly stem from the time- and space-bridging character of telecommunication services (BRUNN, LEINBACH 1991; SHANNON 1997). According to HÄGERSTRAND'S (1970) time geography approach, constraints in space and time have a limiting impact on any form of human interaction. One could consider many deficiencies related to rural households in poor countries as generated by the limited possibilities of interaction between themselves and other agents. This makes information poor, scarce, maldistributed, inefficiently communicated and, hence, extremely costly. Eventually, this has negative implications on the efficiency, productivity, and welfare of all kind of agents in society, e.g. households, firms, and organisations. Through telecommunications, space- and time-bridging constraints and the resulting problems of expensive information flows and limited human interaction could be reduced (GEERTZ 1978; LEFF 1984; SAUNDERS ET AL. 1994).

Based on this argument, the effects one might expect if households use telecommunication services can be grouped according to the benefits that are likely to occur. This means we need to explore how *telecommunication*, on the one hand, substitutes other, more expensive forms of interaction. On the other hand it is necessary to find out whether the increase in the availability of information can augment the households' overall stock of information and consequently lead to better decision making.

Data wise, the demand analysis bases on two field surveys carried out in the Volta Region, Southern Ghana. These surveys mainly assessed household-related variables, information about the respective communities, as well as data related to the communication facilities which provided communication services.

1.4 Study Overview

To give this work a structure that guides the reader from the policy-related institutional and infrastructural topics towards the specific issues of access and use of basic communication services, the following architecture was developed:

In the second chapter, the main characteristics of communication technologies will be linked to the current discussion around the potentials of these technologies to foster economic development. The chapter leads to a synopsis of the various levels of ICTs' economic impact and points at the need for more empirical on-the-ground knowledge.

The third chapter relates global trends of telecommunication sector reform to the current ICT sector developments in Ghana, especially elaborating on institutional issues such as liberalisation, privatisation, and the regulatory environment. Moreover, the specific telecommunication sector situation in Ghana will be analysed and the respective findings shall be related to the other countries in Sub-Saharan-Africa. The question of the extent to which telecommunication infrastructure is currently distributed on the country-level and especially amongst the rural areas of the country will also be tackled in this context. The discussion of the predominant lack of rural telecommunications infrastructure will lead us to the analyses of the available primary data.

Chapter 4 will deal with the supply of and demand for telecommunication services in the selected rural setting. After introducing the major conceptual aspects and hypotheses, the surveying process and the generated data base are explained. In addition to that, the major geographic and socio-economic characteristics of the survey area, namely the Akatsi district, are presented.

In a further progression, the existing telecommunications infrastructure in the survey area will be analysed, followed by a more detailed description of the present public communication facilities. In line with the major research questions, the degree of telecommunication access and use will be characterised and related to the major characteristics of the households under discussion.

In Chapter 5, special attention will be given to the purposes of and the benefits from telecommunication use as they are perceived by the users. The findings will be embedded into reflections on the time- and space-bridging characteristics of telecommunications. Conceptually, this will pave the way to quantify the gains generated by telecommunications if it is used to substitute other forms of communication. The increase in the amount of information flows which is due to the possibility of quicker and cheaper communication, and its impact on the decision making processes of the households will finally be presented in an anecdotal manner.

The last chapter will sum-up the work by reconsidering the major research questions and discuss the results that could be generated. This will lead us to an ex-post evaluation of the conceptual and methodological approaches chosen, to the discussion of policy implications, as well as to the formulation of further research needs.

2 Information and Communication Technologies and Economic Development

Nowadays, the beneficial impact of ICTs on economic growth and development are more or less taken for granted. The complex interdependencies between the technologies and economic mechanisms do not seem to be clear to many people praising the benefits almost axiomatically. With that in mind and in order to embed the research questions raised into the current discussion around those mechanisms, this chapter will first refer to the major characteristics of ICTs. The uniqueness and universality of those characteristics fostered the rapid penetration of all parts of society by the new technologies. The processes involved and their determinating factors will be the next issue addressed.

The various ways in which the implementation of ICTs eventually influence economic organisations, markets, and institutions need to be discussed in a conceptual manner before presenting empirical evidence. Additionally, some specific aspects of ICTs and their impact on regional development and transportation flows are considered.

The chapter closes by relating those somewhat general mechanisms and links to the particular situation of low-income countries and their rural areas respectively, eventually indicating the need for more on-the-ground verification.

2.1 Specification of Information and Communication Technologies

To avoid any misunderstanding, the specific characteristics and particularities of ICTs will now be discussed. It is important to note that we are largely focussing on the capabilities of the technologies related to communication between different economic agents, such as organisations and individuals rather than on the information exchange between machines occurring, for example, when the technologies are directly applied in production processes.

In terms of hardware and software, ICTs can be characterised as technologies which allow the capturing, processing, storing and interactive transmission of information (HAMELINK 1997; HEEKS 1999). This includes common forms of *telecommunication* and also technically more sophisticated Internet applications, but excludes all kind of mass media, i.e. television and radio. Figure 2-1 reflects the different elements of ICTs and highlights one particularly relevant element, i.e. the technological preconditions that lead to the implementation and diffusion of network technologies, applications and services as well as user terminals (hardware).

In this context, we refer to the fundamental improvements occurring with the transition from analogue to digital technology, which was a crucial factor in fostering the convergence of the different technologies. These days, if one has access to a telephone line and a computer, nothing more is required for all kinds of ICT applications, i.e. voice telephony, fax-transmission, electronic mail, and the use of the WWW.

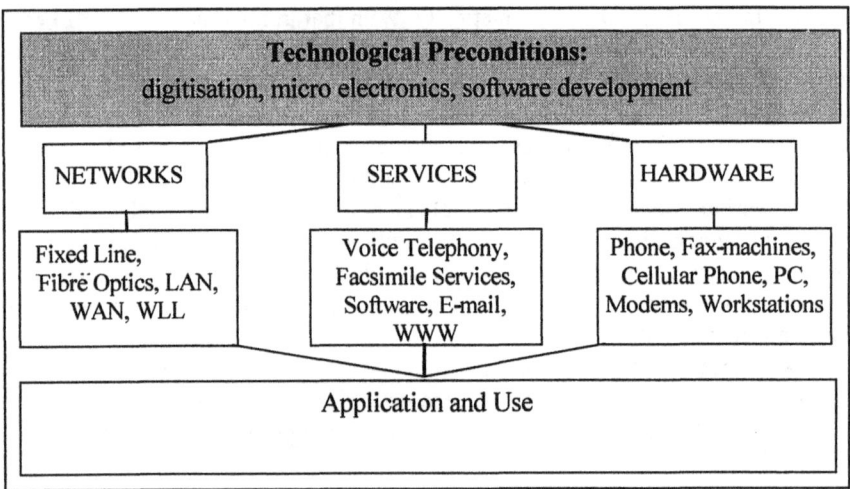

Figure 2-1: Technical elements of ICTs (Source: BERTOLINI 1998)[2]

However, it is not the technology as such that initiated so much change in communication processes, but rather the impact of ICTs on the universe of information and its transmission through the following essential characteristics (cf. BEDI 1999; BRUNN, LEINBACH 1991; POHJOLA 1998).

---

2   For the remainder of the study, sources of figures and tables are classified as:
     Source if the cited document does already contain a figure or table similar to the one produced;
     Data source if the data from the source was used to design a table or figure;
     Cf. is used if ideas and thoughts expressed in the original document were summarised in tabular or graphic form;
     if *no source is indicated*, data as well as design of the table or figure base on own data and concepts.

1) ICTs allow the decoupling of information from its physical repository over increasing geographical scales with declining unit costs. In other words, they allow the immediate transmission of information independently from physical movement, a feature that is often referred to as the space- and time bridging characteristics of ICTs.
2) The use of the technologies is non-rival. This simply means that the increase in the overall number of users generates benefits for all other users of the network. These so-called network externalities occur with each added subscriber: any earlier subscriber could potentially exchange information with the new member of the network.
3) ICTs can be applied in any sector of the economy. This pervasiveness can be derived from the digital mode underlying the technologies. One important standard for non-verbal communication would, in this sense, be the Internet Transmission Protocol TCP/IP (SEIBEL ET AL. 1999). For verbal communication, the common language used by the communicating persons can also be regarded as an underlying standard.

2.2 The Diffusion and Implementation of ICTs

This paper has, thus far, showed the unique characteristics of these technologies. Their rapid dissemination is due to those characteristics that "have greatly reduced the cost of transmitting and processing information, altered the cost structures of [...] industries, created new ways of meeting a wider range of communication needs at lower cost [...] and increasingly integrate information and telecommunications technologies and services" (SAUNDERS ET AL. 1994, p. 122). In doing so, new ICTs were said to be a key to maintain and develop the competitiveness of countries, regions, and companies. They increasingly became an integral part of financial services, commodities markets, media, and transportation. Furthermore, they provide vital links between manufacturers, wholesalers, and retailers and enable the former to flexibly reconfigure and manage their supply networks as the corporate objectives change (SAUNDERS ET AL. 1994).

All these aspects make clear that ICTs and the global economy go hand in hand and that "countries and firms that lack access to modern systems of telecommunications cannot effectively participate in the global economy" (SAUNDERS ET AL. 1994, p. 122).

To understand the degree to which the technologies are implemented in a specific country and, therefore, the extent to which it is able to benefit from ICTs, it is crucial to recognise the major diffusion mechanisms. These elements – namely the determinants and the dimension of diffusion processes – are now illustrated in a rather pragmatic manner without analysing inherent processes, such as innovation processes, communication channels or time-related and social circumstances (cf. ROGERS 1995).

In order to conceptualise the dimensions and determinants of the diffusion processes, we mainly refer to the framework developed by THE MOSAIC GROUP (1998) and to HANNA, GUY, ARNOLD (1995). The extent to which the technologies are embedded within society will form another cornerstone of this section which eventually will lead us to a discussion of the economic benefits we can expect from these technologies (PRESS ET AL. 1998).

2.2.1 Determinants and Dimensions of ICT Diffusion

2.2.1.1 Determinants

A country's ICT capabilities are closely linked to determinants of technology diffusion. This allows the conclusion that any further development of the sector will be a result of these determinants. By selecting the most important factors that determine the diffusion of ICTs on the level of individual countries, one should look more closely at the role of governments and their policy orientation towards global competition and market integration. This orientation, again, depends on country-specific factors such as its factor endowment, the effectiveness of constituent groups and the existence of ICT-related production industries.

The Role of Governments

Sectoral policies became an important means of promoting the diffusion of ICTs. The implementation of such policies was initially started by the OECD countries but increasingly became a universal feature. It goes beyond the scope of this work to elaborate on these developments in detail. Nevertheless, an analytical framework set up by GUY and ARNOLD (1987) will be applied to characterise different ICT-policies. According to this framework, one can distinguish between three policy-related dimensions.

The first dimension distinguishes government activities that are either ICT-*specific* or ICT-*related*. ICT-specific measures aim at promoting the production and use of ICT goods and services by individuals and firms. A relevant example would be the provision of grants for a telecommunication demand analysis in rural areas. ICT-related policies, however, do not specifically focus on ICTs "but nonetheless influence its generation and use. They include [...] competition policy, financial and trade liberalization" (HANNA, GUY, ARNOLD 1995, p. 39).

Secondly, and according to the degree of governmental direction, one can identify the so-called *hands-on* and *hands-off* policies. The former should encourage firms and individuals to follow goals that are in line with the government's strategic priorities, e.g. the allocation of loans or subsidies for ICT-related research and development (R&D) projects. The latter do not have specific foci and could also affect other sectors.

Thirdly, policies can be set up and developed to either initiate and increase the production of ICT-related services and goods, and to enhance their use, or to build the link between producers and users. Classically, this means that governments stimulate technological activities and promote the indigenous generation of ICTs. Diffusion policies are then encouraging the acquisition and application of the technologies by firms and individuals. Means to bridge users and producers are regarded as crucial to speed up the adaptation process and to more closely link the users to the ongoing technological developments.

Figure 2-2 provides a synopsis of the policy-related dimensions listed above and provides examples of some of the aforementioned instruments.

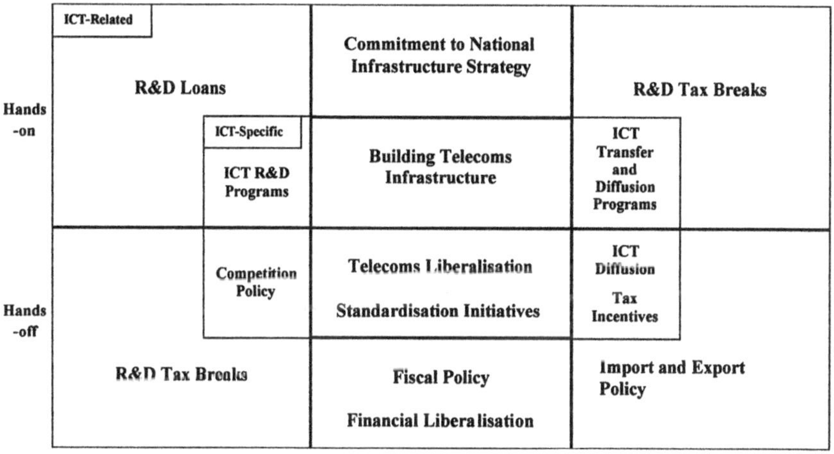

	ICT-Related			
Hands -on	R&D Loans		Commitment to National Infrastructure Strategy	R&D Tax Breaks
		ICT-Specific ICT R&D Programs	Building Telecoms Infrastructure	ICT Transfer and Diffusion Programs
Hands -off		Competition Policy	Telecoms Liberalisation Standardisation Initiatives	ICT Diffusion Tax Incentives
	R&D Tax Breaks		Fiscal Policy Financial Liberalisation	Import and Export Policy

Figure 2-2: Matrix of ICT-related government policies (Source: HANNA, GUY, ARNOLD 1995)

Country-Specific Determinants

Government action, as classified above, has an impact on most other determinants of ICT diffusion. According to PORTER (1990), the general determinants of a country's competitive advantage can be grouped into various divisions. The three that are particularly important for the diffusion of ICTs will now be explained.

1) Factor conditions are those determinants that primarily concern the factors of production, in our case the inputs for the providers of ICT services. This sounds very technical if only infrastructural, technological, financial, and human resources are borne in mind. It becomes more complex, however, if one considers aspects of culture, geography, and available information resources (HANNA, GUY, ARNOLD 1995; THE MOSAIC GROUP 1998). Table 2-1 shows factor conditions that, on the one hand, are subject to change due to outside influences, e.g. government action. On the other hand, the table shows that natural factors may well be unchangeable and form obstacles that may result in serious disadvantages for a country's infrastructural development, for example. In this context, one can recognise the advantage of a flat landscape morphology for the installation of wireless technologies. The installation of wireless systems in an hilly area clearly results in higher installation prices and, hence, a slower diffusion process, provided financial constraints are prevalent.

Factors	Example
Culture	- Language
Human resources	- Education - Skill levels
Financial resources	- Liquid assets - Investment sources - Obstacles to flows of capital
Infrastructure	- Domestic telecommunications infrastructure - International telecommunications
Information resources	- Availability of technical information - Free flow of foreign information
Natural conditions	- Obstacles to infrastructure development

Table 2-1: Factor conditions and selected examples of ICT diffusion (cf. THE MOSAIC GROUP 1998)

2) Like any other technology or innovation, ICTs are spreading due to an existing demand. This demand emanates from private individuals, businesses, civil society, and the public sector. The activity of those constituent groups and their status in society eventually determines whether efforts aimed at spreading ICTs will be successful.

3) The strategic and structural state of the ICT market is, according to THE MOSAIC GROUP (1998) and NELSON (1993), an element that either fosters or hampers the diffusion of all kinds of ICTs. The condition of the market is reflected by the competitive environment and innovations. The former mainly comprises of the existence and strength of market entry barriers, the costs and prices of ICT-related goods and services ,and the degree of market concentration. The latter refers to a country's innovation environment and, hence, its educational system as well as to its R&D capacity (MOWERY, OXLEY 1995).

2.2.1.2 Dimensions

The dimensions of ICT diffusion more or less reflect the availability of ICTs, i.e. the extent to which the technologies are dispersed, and the spatial and sectoral pattern of this dispersion. Moreover, structural dimensions reflect the degree to which ICT applications have become an integral part of a country's social and economic matrix.

Looking at *pervasiveness* in a quantitative manner, one could simply consider the percentage of users of specific ICTs, e.g. telephones, fax, or email services within the society. The most commonly used indicators in this respect are the *teledensity*, i.e. the number of telephone main lines per 100 inhabitants, or the number of Internet users per 1.000 citizens, etc.

Spatial dispersion describes the physical availability and spread of ICTs within a country. Typically, one should expect the spatial dispersion over time to correlate with the settlement pattern of a country and the degree of centrality of the respective settlements. In this sense, *spatial pervasiveness* will be reached if the services are at least potentially available in all communities of a specific country, regardless of their size or their remoteness.

Sectoral dispersion reflects the penetration of ICTs within the major segments of modern societies, namely the private sector, the public sector, and actors from civil society.

A country's *connectivity infrastructure* consists of different elements, such as the state and capacity of the domestic telecommunication network, the capacity of its international gateway, as well as the sophistication of the hardware available to the users (cf. Figure 2-1).

Features such as the degree of competitiveness and the number of providers within the different market segments can be used to reflect the *organisational infrastructure*. It is generally believed that "an open, competitive market with low barriers to market entry is more conducive to high rates of take-up by subscribers, wider proliferation of the physical infrastructure, and the provision of a wider variety of access" (THE MOSAIC GROUP 1998, p. 4).

The last dimension of ICT diffusion is the *degree and sophistication of use*, i.e. the extent to what the application of the technology has really caught hold within a country and became an integral part of that country's socio-economic system. This dimension therefore reflects the range of applications used and the intensity of this usage. This demand-centred point of view will especially be important in Chapters 4 and 5 of this research.

A more integrated view of the explicit determinants and dimensions of ICT use is provided by HEEKS (1999). The author's systemic view of ICTs reflects how the technology and the transferred information are allocated within societies' information systems and their institutional and factorial environment. According to HEEKS (1999, p. 3), this requires the consideration "of processes of purposeful activity and people to undertake those processes".

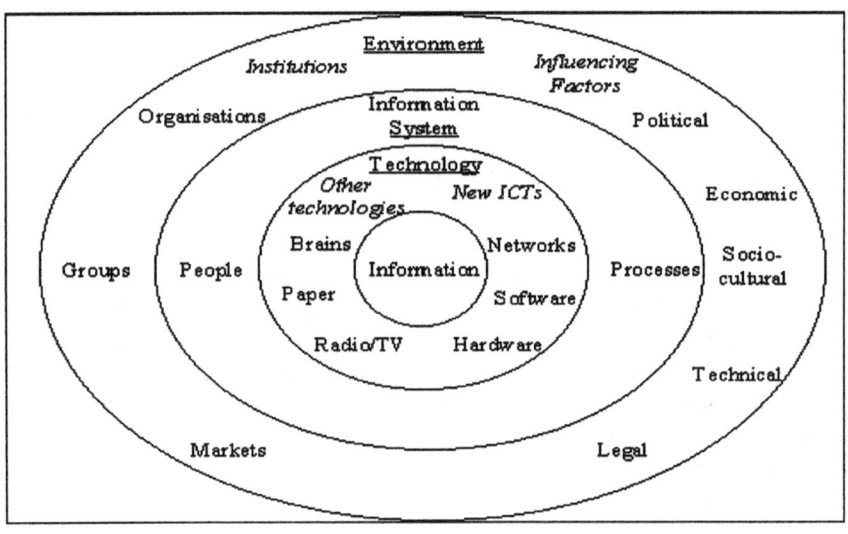

Figure 2-3: Systemic view of ICTs within modern societies (Source: HEEKS 1999)

As reflected in Figure 2-3, people, processes and the means to transmit information are building the information system, which, again, cannot be viewed in an isolated manner. The information system itself exists within and is determined by *environmental conditions*. Those mainly consist of organisations, groups, and markets on the one hand and socio-economic, political, technical and legal factors on the other.

2.3 ICTs and Economic Development – Concepts and Empirical Evidence

Presuming a certain degree of ICT diffusion, this diffusion will lead to significant changes to the economic environment (TALERO, GAUDETTE 1995). On the one hand, the production of information and communication technologies and services becomes more and more important and – in some countries such as the USA or Finland – has exceeded the production of the traditional industries. On the other hand, the successful application of ICTs results in changes for economic agents, organisations and markets. According to many authors these changes usually have a positive impact on the overall economic situation (cf. BEDI 1999; HANNA, GUY, ARNOLD 1995; POHJOLA 1998; TALERO, GAUDETTE 1995). The aim of this section is to confront a priori views and theories about the impact of ICTs on economic development and growth with the empirical proof for those views. This will be done by reviewing major theoretical aspects and then sketching related empirical information. We will do so by distinguishing between the macro- and the microeconomic level. Finally, the potential impacts ICTs have on regional development and transport will be described.

2.3.1 Theoretical Aspects

The acquisition of information accrues costs. This is the same everywhere and applies to any kind of economic activity. Depending on its costs, the amount of information available is subject to change. Accordingly, information shortages are often due to the high prices to retrieve information. A result of those high prices is an increase in uncertainty about markets which has negative implications for the participating agents and institutions (cf. AKERLOF 1970; BEDI 1999; STIGLITZ 1989).

As mentioned earlier, the main feature of ICTs is their ability to distribute and receive all types of information in a cost-effective and non-rival manner (cf. 2.1). Thus, "the key role of ICTs is that they [are tools that] may be used to acquire and process information and reduce uncertainty" (BEDI 1999, p. 8) at high speed and low cost.

This feature leads LEFF (1984) to the conclusion that ICTs lower transaction costs by
- reducing time and monetary costs of transmitting information,
- having a negative impact on resource allocation decision costs between different sectors and agents of the economy,
- leading to an increase in search activity and raising the quantity and quality of available information, enabling marginally better decisions,

- enabling immediate interactive communication and negotiation, which not only includes verbal conversation but also the transmission of data.

Lower transaction costs are manifesting themselves in all parts of a society. For this work, their impact on organisations and markets is of specific interest.

Organisations

For households, firms, and the public sector alike, the reduction of transaction costs lowers the price of a piece of information and one should expect an overall increase of the amount of information transmitted (cf. Figure 2-4). The ease with which information can be acquired, potentially, also increases the quality of the information available to organisations and individuals. Old and obsolete information may be replaced by up-to-date and, thus, more complete and reliable information. Provided the agents are taking advantage of these potentials, an overall improvement of the decision-making process will be the case.

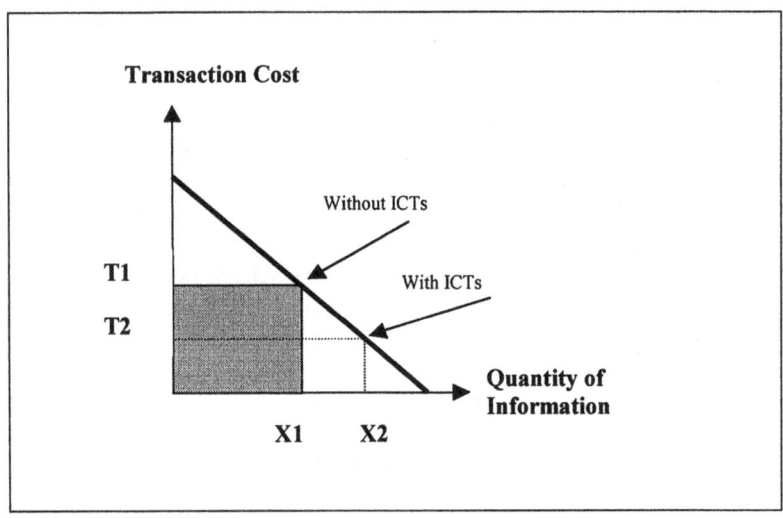

Figure 2-4: Impact of ICTs on the reduction of information transaction costs (cf. LEFF 1984; VARIAN 1987)

Markets

The ability of ICTs to increase search activities and eventually raise the quantity and quality of available information reduces market uncertainty and should, in turn, enhance participation in factor and product markets throughout all geographic scales (BEDI 1999; LEFF 1984; NORTON 1992). Concerning the emergence of markets it is important to note that – in the case of very high transaction costs – market failure occurs. But again, the reduction of information and negotiation costs through "the presence of ICTs may be expected to lower the threshold that needs to be overcome for markets to emerge" (BEDI 1999, p. 11)(cf. Section 5.2).

Negatives

From the conceptual point of view, there are also potential negative effects. BEDI (1999) and LEFF (1984) raise the following points.

A fear that especially became an issue in the developed world is the effect of ICTs on employment and income distribution. The effect on the former might clearly be generated by the automation of work processes by ICTs as they substitute human labour. Concerning the latter, ROCHE and BLAINE (1996) see a shift of demand for labour requiring traditional skills to those of specialised and technical skills as one potential source of danger. So far there are no general conclusions available. Empirical studies indicate, however, that both negative and positive impacts of ICT diffusion on employment figures and income inequalities could be measured (cf. BEDI 1999).

As the distribution and use of ICTs is interdependent on existing income and wealth distribution one could conclude – especially for countries with lower incomes – that access to ICTs will remain restricted to the wealthy parts of the society and, in consequence, worsen information and income inequalities (BEDI 1999; SAUNDERS ET AL. 1994).

2.3.2 Empirical Evidence

It was argued earlier that ICTs as cross-cutting technologies might have an impact on all kinds of organisations, institutions and economic agents and – through these – foster the spread of markets and increase welfare and productivity. We will now confront those conceptual views with empirical evidence by reviewing papers that consider both, the macro-economic and the micro-economic level.

2.3.2.1 Macroeconomic Evidence

On the macro-economic level one has to distinguish between at least four different approaches to link economic development and growth to ICTs.

1) The first approach focuses on the link between the intensity of ICT production and investment, and the economic performance reflected by GDP measures. Clearly, economies have benefited from the production of and investments in ICTs; be it directly through the production of telecommunications and information technology hardware and services, or indirectly through multiplier effects, such as the overall increase of internal consumption that is due to the positive ICT sector developments (HEUERMANN 1999).

2) Secondly, ICT production and use lead to growth effects generated by tax income. HEUERMANN (1999) points out that in 1995 about 5.7% of all revenues of the biggest telecommunication providers went into the public budgets of the OECD countries, not including the indirect incomes generated from the value-added and turnover tax paid. Another income source for the public hand derives from the tendency to privatise national telecommunication operators and the licensing of radio frequencies on which mobile cellular technologies operate.

3) Using cross-country time-series data, NORTON (1992) tries to verify whether the existence of telecommunication infrastructure reduces transaction costs and consequently generates an increase in output. Without replicating the methodology and econometric aspects, the results indicate that there is a positive relationship between the existence of telecommunication infrastructure and economic growth[3]. The same can be observed in Figure 2-5 which shows the relationship between GDP per capita and the number of telephone lines per 100 inhabitants of more than 150 countries for the year 1995.

4) Concerning information technology markets, POHJOLA (1998, p. 8) points out that the "size of the market seems to correlate positively with the standard of living as measured by GDP per capita in purchasing power parities". He recommends caution, however, concerning the interpretation of causality: "It may be more likely that high income countries tend to purchase more infor-

---

3　Telecommunication infrastructure is represented by the proxy variables telephone density at the beginning of the time series observation and the average telephone density over the time period of the sample. Economic growth is reflected by a vector of variables including per capita income, mean annual population growth, mean money-supply growth, mean growth in the ratio of government spending to output, mean growth in exports as a proportion of output and mean growth in the rate of inflation (BEDI 1999; NORTON 1992).

mation technology rather than that high spending in information technology makes countries rich" (POHJOLA 1998, p. 8).

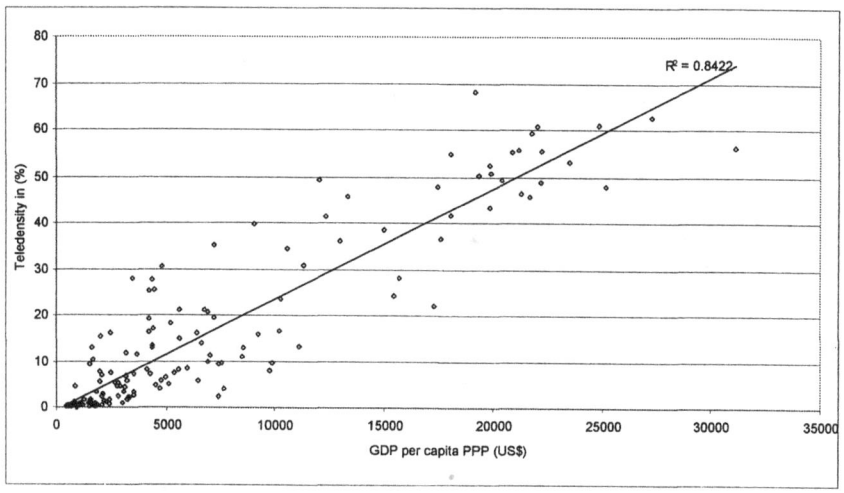

Figure 2-5: Correlation between GDP per capita (PPP) and the number of telephones per 100 inhabitants for 153 countries in the year 1995 (Source: HEUERMANN 1999)

Based on those experiences, the problems of showing a clear dimension and causality of the impact of ICTs on economic growth are obvious. BEDI (1999, p.18) argues that this is mainly due to "the variety of factors that may be responsible for growth and the endogeneity of ICTs and output". He also argues that regardless "of the causal linkages, it is clear that there is a positive association between ICTs and growth" (BEDI 1999, p.18). Similarly, SAUNDERS ET AL. (1994, p. 157) refer to the complex relationship between ICT investment and economic activity and state that "the benefits of a specific telecommunications project [...] cannot readily be identified and measured by the aggregate international comparison of input output tables, by the analysis of relations between GNP and telephone availability or usage". To address those shortcomings, they suggest focusing on the micro-economic assessment of the benefits that are related to ICTs.

2.3.2.2 Microeconomic Evidence

We will now review some empirical aspects of the extent to which the technologies have an impact on the microeconomic level, concentrating on enterprise productivity, market development, and consumer welfare.

Productivity Gains

As described above, the application of ICTs should speed up and enhance decision-making processes by reducing the cost of exchanging and generating information. For all kinds of enterprises the application should thus lead to an increase in productivity. On an aggregate level, this does not necessarily hold true if empirical research is considered (cf. BAILY 1986; ROACH 1991). According to the investigation the authors carried out in the 80s "the US economy failed to show productivity gains expected from the investment in information technology" (BEDI 1999, p. 19). In fact, the rapid increase of IT application is opposed by a slow down of overall productivity, or according to Solow: "You can see the computer age everywhere but in the productivity statistics" (THE ECONOMIST 2000, p. 13). There are several other studies that discuss this *productivity paradox* (cf. LOVEMAN 1994; MORRISON, BERNDT 1990).

A later approach, developed by BRYNJOLFSSON and HITT (1996), based on an investigation of more than 350 large enterprises in the late 80s and early 90s, became an important example that reconciled the aforementioned paradox. The authors managed to empirically prove a positive link between output elasticity and the ICT-related capital stock and staff (BEDI 1999). The authors themselves explain the discrepancy towards the older work. They state that their data set might be more accurate, as it uses very detailed company data, but also suggest that the time period covered by their sample does make a difference: the "more recent data incorporates learning effects and changes in business processes that have been instituted in response to the advent of computers" (BEDI 1999, p. 19). Learning from those studies, there is no clear evidence that ICT investment will lead to an increase in enterprise productivity. However, one could suspect that there is a substantial time lag between ICT investments and their potential benefits.

Market Spread and Institutional Arrangements

The empirical evidence concerning the impact of ICTs on the spread of product and factor markets is so far rather limited and primarily of anecdotal nature. BERTOLINI (1998) referred to a case in which a small mechanical engineering company in Southern Germany was able to strategically position its products on the South-east Asian market by co-operating with an Australian partner. The use of ICTs was crucial not only to find an adequate partner, but also to initiate and maintain the partnership over the given geographical distance. From the developing world there are reports of farmers that use ICTs to obtain price information about the products in order to better negotiate with the traders and eventually increase their profits (BAYES, VON BRAUN 1999; BERTOLINI ET AL. 2000).

Consumer Surplus

Considering the demand for information and communication services from a consumptive rather than a productive perspective, one should draw attention to the consumer welfare concept. Providing empirical examples of consumer surplus estimation, SAUNDERS ET AL. (1994) in particular refer to two different modes of measurement.

The first mode requires the observation of the response of users to a change in price for the services. In the given example, they calculated the consumer surplus, estimating the price elasticity of demand for telephone call traffic by observing the change in traffic associated with a change in the call charges. Building upon this, one is able to estimate a part of the demand curve for the use of telecommunication services. Through the comparison of price increase information, the consumer surplus can be measured in an economy that, for instance, is affected by price inflation (cf. Figure 2-6). This was carried out by the authors in El Salvador in the late 70s looking at panel data from 1964 to 1982. In the time period between the beginning of the observation and 1977 there were no significant tariff changes for telecommunication services but domestic consumer prices increased by around 80%. This resulted in a consumer surplus of which a portion could be estimated by tabulating the prices that consumers had been willing to pay when they acquired telephone service (SAUNDERS ET AL. 1994).

The consumer surplus can alternatively be measured by estimating how much the users gain by using ICTs rather than an alternative way of communication. The gain derives from the fact that the one or the other alternative is more costly in monetary or time-related terms. Cost savings occur, for instance, if home banking replaced the physical visit to the bank counter. On top of the time saved, one might save money through the reduction of transport costs. For some developing countries, SAUNDERS ET AL. (1994) present examples in which the surplus from telecommunication service utilisation was generated by the fact that the users – in the absence of the services – would have used more cost and time intensive means of communicating, such as postal services, physical travel, or sending a third person (cf. Chapter 4 and 5).

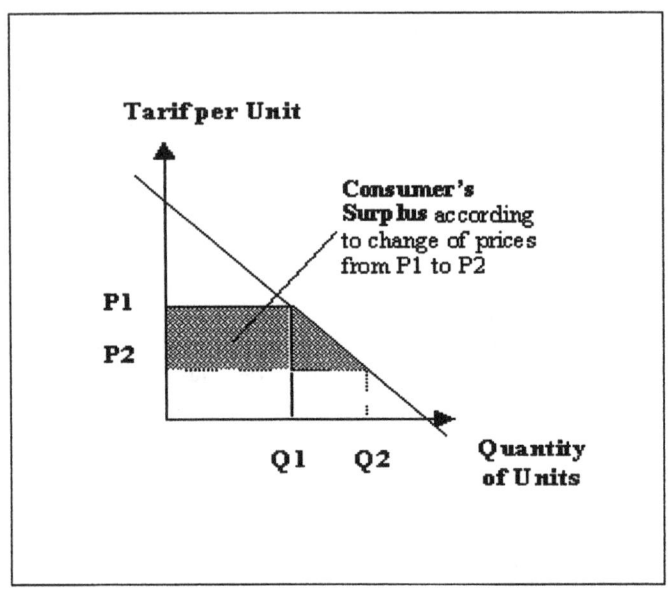

Figure 2-6: Consumer surplus calculated on the basis of price changes (cf. Varian 1987)

2.3.3 Issues of Regional Development and Transportation

In the forthcoming section, insights into the impact ICTs might have on issues of centrality and regional development as well as on physical movement will be presented.

2.3.3.1 ICTs, the Central Place System, and Regional Development

The linkages between communication technologies and the spatial order of settlements, as well as the relationship between urban centres and their hinterlands were analysed by CHRISTALLER as early as 1933 (cf. CHRISTALLER 1966). By "pursuing the hypothesis that a hierarchy of places depends on the relative size and economic importance of central places [he] used the number of telephones as a proxy for a common measure of centrality in his examination of the size, number and distribution of central places" (SAUNDERS ET AL. 1994, p. 122). Although the number of telephones might nowadays not be an adequate measure for centrality in most developed countries, KILGOUR (1982) argues – based on research carried out in Costa Rica – that the telephone density could still be used as a variable to reflect the system of the settlement hierarchy in Christaller's sense.

Apart from this hierarchy, SAUNDERS ET AL. (1994) quote examples from central Chile, which show the relationship between larger cities and their hinterland, comprising of smaller towns and villages. Results indicate that up to 90% of the region's smaller towns exchanged the clear majority of their calls with urban centres. If one presumes that there is more economic activity and specialisation in the latter, one could conclude that the analysis of telecommunication traffic pretty much reflects economic structures and interdependencies on the sub-national and, hence, the regional level (WEBBER 1980).

It might be interesting to utilise these findings within the context of regional development. The concrete implications are the extent to which telecommunication services influence locational decision-making and therefore foster regional balance.

There is a long tradition in regional science that aims at avoiding undesirable spatial concentration of modern sector activity (RITTER 1998). On the one hand this leads to negative agglomeration effects such as traffic congestion with all its negative externalities[4] and other shortcomings, e.g. the problems of keeping pace with the provision of public goods and infrastructure. On the other hand, inequality and polarisation between urban settlements and rural and remote areas could develop to an undesirable extent. One means of reducing those negatives is to strengthen the economic and administrative position of settlements within the central place system that are located outside the urban agglomerations. ICTs are seen as an important means of supporting this process and, in fact a lot of thinking was invested in the idea that ICTs particularly benefit remote and less developed regions (cf. EUROPEAN COMMISSION 1996; TETSCH 1985).

The problem of rural-urban migration is closely related to issues of regional development. One could conclude that, in line with the issues mentioned above, the spread and use of ICTs can be seen as a way of reducing undesired migration due to their potential to increase economic productivity in rural areas by, for instance (cf. SAUNDERS ET AL. 1994):

- speeding up the information flow with regard to medical needs,
- offering ways of accessing the latest information about input and output prices or new agricultural techniques,
- enhancing the use of transport and movement of supplies and marketables.

---

4   I.e. high energy consumption, environmental effects, inefficient use of time are mentioned (cf. RENAUD 1981).

However, overall regional development is a long-term process and therefore rural-urban migration would not be stopped in the short run by just providing communication services to rural areas. The contrary might even be the case "[...] since the migrants could more easily keep in touch with their home village and family and would not have to endure long periods without access to local information" (SAUNDERS ET AL. 1994, p. 129).

2.3.3.2 ICTs and Transportation

There are strong linkages between the ICT sector and other means of communication and transport, such as roads, railways and the postal system (ITU 1988; TYLER 1978). Focusing on the relationship between physical transport and ICTs, at least two dimensions that potentially have positive effects on development and growth need to be considered. The first relates to the extent to which ICTs and transport complement each other, whereas the second dimension reflects the potential of ICTs as a substitute for transport.

Complementarity

In industrialised and industrialising countries ICTs became an everyday means to improve the efficiency of logistic firms and other transport-related enterprises. In this context it is apparent "that benefits can be reaped at various levels of technology, including a one-vehicle truck business using a public telephone to locate a destination or secure a return load and a large business using a sophisticated system of radios to locate and identify vehicles automatically and transmit posting instructions from a central control location" (SAUNDERS ET AL. 1994, p. 137).

Through these measures, the communication channels are opened that enable the specific entrepreneur to optimise the use of their fleet more efficiently. Empirical evidence from the United States and the Soviet Union, for instance, shows that this entails full loading and reduces idle time, misrouting and empty return journeys (ITU 1988; LATHEY 1975; PYE TELECOMMUNICATIONS 1976). This allows the enterprise not only to be more competitive and decrease prices offered to the customer but also to react better to the latter's needs and requests.

Substitution

Be it just telephony or videoconferencing via fibre optic connections, when it comes to information and communication services, there is a long tradition of discussing the end of personal meetings and business travel (cf. FRAUNHOFER GESELLSCHAFT 1995).

Research in this field is generally scarce. Research that has been carried out mainly concerns business-related traffic and analyses whether or not ICTs can substitute personal meetings to such a degree that the same results, i.e. user satisfaction, could be achieved (JOHANSEN, VALLEE, SPANGLER 1979). The results of these conceptual studies show "that a partial substitutability of telecommunications and travel seems to be feasible when a full range of costs and behavioural factors are considered[5]" (SAUNDERS ET AL. 1994, p. 144).

Experiences from the field are mostly restricted to the substitution of business trips by the use of teleconferencing in large organisations, such as ARCO, Boeing, IBM, NASA, or Procter and Gamble. According to SATELLITE COMMUNICATION SERVICES (1981), IBM could, for instance, incur cost savings due to teleconferencing of US\$ 414,000 in 1979 and US\$ 830,000 in 1980 respectively.

Only few studies provide evidence related to telecommunication travel substitution in the South. CLEEVELY and WALSHAM (1980) surveyed two regions in Kenya, one farther away from Nairobi and more sparsely populated, another one, more densely populated, and closer to the capital. The results show that the telephone users in the first region make about 80% more use of the facilities than the users in the second. From this fact, the researchers concluded that "some relative substitution of telephones for travel was taking place in the more sparsely populated and distant [...] region" (SAUNDERS ET AL. 1994, p. 146). CHU; SRIVISAL and SUPHAILOKE (1985) more directly found out that over 60% of office workers that were interviewed in a survey of telephone users in four districts in Thailand would have made personal visits to carry out their tasks if they not had the possibility of using telecommunication services. Other alternatives were to send messages through a third person (17%) or through the postal service (11%).

Travel Generation

In addition to the complementary and substitutional impact, better communication and information systems may also induce physical movement. Although there is very limited evidence and research on these issues, it seems rather plausible to assume that the ICT-borne rise in market participation and the emergence of markets respectively (cf. 2.3.1) should result in an increased exchange of goods and services. In turn, a higher demand for face-to-face negotiations and

---

5   The latter especially occur in the case of teleworking or telecommuting, which faces "internal resistance to nontraditional forms of organization and management within large companies and [...] by external criticism from researchers and labor unions that telecommuting makes workers vulnerable to economic exploitation" (SAUNDERS ET AL. 1994, p. 144).

decision-making processes occurs. Thus, transport and travel events would be generated (FRAUNHOFER GESELLSCHAFT 1995). Similarly, the aforementioned customer friendliness and the decrease in prices that is due to the complementary effects of ICTs on transportation should eventually lead to a higher demand for transportation services and, hence, increase physical movement. Moreover, the possibility of direct communication between people might result in a higher demand for face-to-face communication. A survey by CLARK and UNWIN (1981) – covering rural-urban communication flows in South England – relates to this phenomenon, indicating that telecommunication "did not and could not acceptably replace the face-to-face contacts upon most [...] relations were based" (SAUNDERS ET AL. 1994, p. 146).

2.4 Implications for Low-income Countries

While considering ICTs and economic growth in a rather general way, the overall subject of this dissertation thesis makes it necessary to focus on the specific issues low-income countries[6] have to deal with when it comes to ICTs. In this respect, the first part of the next section sums up the views shared by most international institutions on the potential role of ICTs in enabling the countries to meet more effectively their development goals (cf. G8 2000). We will confront these views with infrastructural facts from the sector, mainly comparing ICT-related indicators from developed and low-income countries.

This discussion leads to a further intention of this section: down-scaling the focus towards rural areas in low-income countries and their specific problems related to ICTs. Judging from the existing literature, political viewpoints, and secondary data we eventually recognise the need for empirical analysis that will be addressed by the remainder of the work.

2.4.1 The Digital Divide versus the ICTs' Leapfrogging Potentials

In the past, the causal relationship between access to telephone service and economic benefits was only rarely seen as beneficial for development. This led to the fact that fostering the diffusion of telecommunication services did not become a pressing policy and investment issue. With the convergence of information tech-

---

6   The term 'low-income country' will be understood according to the definition provided by the World Development Report 1999/2000 (WORLD BANK 2000) and used synonymously with the term *developing country*.

nology and telecommunication applications, especially in the form of the Internet, this changed dramatically:

"The changing role of telecom services in all economies, and an expanding research literature documenting the barriers, restrictions, costs and penalties of inadequate telecom service for participation in economic and social life in all countries have brought the issue of telecom development to the forefront of policy and investment analysis" (MELODY 2000; p. 635).

In light of this, most industrialised countries started to develop their information infrastructure to an extent that allows them to keep track with economic and technical innovations and prepare for e-commerce, m-commerce and business-to-business applications.

Whether all these developments eventually have an economic impact far greater than that of any other past technological revolution or whether it is nothing more than a stock market bubble is open to discussion (THE ECONOMIST 2000). There is no doubt, however, that ICTs have an impact on the global economy to the extent that they reinforce the globalisation process:

"By reducing the cost of information and communication, IT has helped to globalise production and capital markets. In turn, globalisation amplifies the economic gains from IT" (THE ECONOMIST 2000, p. 43).

Considering these macro-economic gains, there is a growing concern that developing economies will be left behind and the already existing income gap between the developing and the industrialised world will widen further due to the *digital divide* (BEDI 1999; THE ECONOMIST 2000; SEIBEL ET AL. 1999; WORLD BANK 1998). This divide becomes obvious if one considers the fact that "the rich countries account for only 15% of the world's population but 90% of global IT spending and 80% of Internet users" (THE ECONOMIST 2000, p. 38). This picture gets even more disadvantageous if one weighs ICT-related infrastructural indicators of OECD countries against those of low-income countries (G8 2000). Figure 2-7, in this respect, indicates that the number of computers and mobile phones per 1.000 inhabitants does significantly differ between the richest and the poorest countries. Furthermore, there are – on average – around 25 times more telephones and about 50 times more fax machines per 1.000 inhabitants in OECD as compared to the low-income countries (cf. Figure 2-7).

Yet again, to only consider the differences in infrastructural endowment is too short-sighted. The indirect effects that occur due to the lack of available information (cf. Section 2.3) are adding to the existing economic shortcomings in poor countries. The macro-economic consequences, i.e. the costs of a lack of access can be summarised by pointing out the following three issues (cf. Section 2.3.2.1

and BEDI 1999; THE ECONOMIST 2000; TALERO, GAUDETTE 1995; UNDP 1998; WORLD BANK 1998).

First, the lack of ICTs results in a lack of turnover and economic benefits generated by their production and use. Apart from these direct effects, the aforementioned multiplier effects (cf. Section 2.3.2.1) do not occur and, on top of that, the lack of home production of ICT equipment must lead to imports that require foreign currency (TALERO, GAUDETTE 1995; UNDP 1998). It is therefore not too surprising that a computer in Bangladesh costs the equivalent of eight years' average salary (THE ECONOMIST 2000).

Second, tax revenues derived from the production and use as well the indirect effects of ICTs can only be realised in a very limited manner.

Third, the macro economic manifestation of the micro-economic potentials of ICTs in terms of enterprise productivity, market spread, and consumer surplus as well as issues of regional development and increased transport efficiency does not apply if ICTs are only developed marginally. This has implications for the competitiveness of the country and its participation in global markets. For instance, the fact that multinational firms have an advantage from the network externalities of ICT applications will enable them to establish their already dominant position at the expense of local firms that are late adopters of the technologies and therefore inevitably fall behind in terms of competitiveness. Indirectly, one may therefore suspect that there are also negative effects on the overall economic situation, including innovativeness and employment (cf. Section 2.3.2; HANNA, GUY, ARNOLD 1995; TALERO, GAUDETTE 1995; UNDP 1998; WORLD BANK 1998).

International support form the World Bank, the UNDP, USAID and other bilateral and multilateral donors[7], as well as legal and regulatory commitments by national governments led to the situation that, nowadays, all countries of the world, are connected to the global information infrastructure. The connectivity map of Africa (cf. Figure 2-8), for instance, shows the impressive improvements in both connectivity and bandwidth of the international gateway during the last couple of years (ITU 1998; JENSEN 2000; WORLD BANK 1998).

---

[7]   For a comprehensive overview of the funding activities of international aid agencies and international institutions cf. WOYTEK., R. (1997): Schlußbericht. Zugang zu Informationen für Nichtregierungsorganisa-tionen. Ein Vergleich internationaler Initiativen, GTZ, GATE, ISAT. Bad Homburg, mimeo; and MÜLLER-FALCKE, D. (1999): Informations- und Kommunikationstechnologien in Entwicklungsländern und Initiativen zu deren Förderung - Ein Überblick; Studie im Auftrag der Gesellschaft für Technische Zusammenarbeit GmbH (GTZ), mimeo.

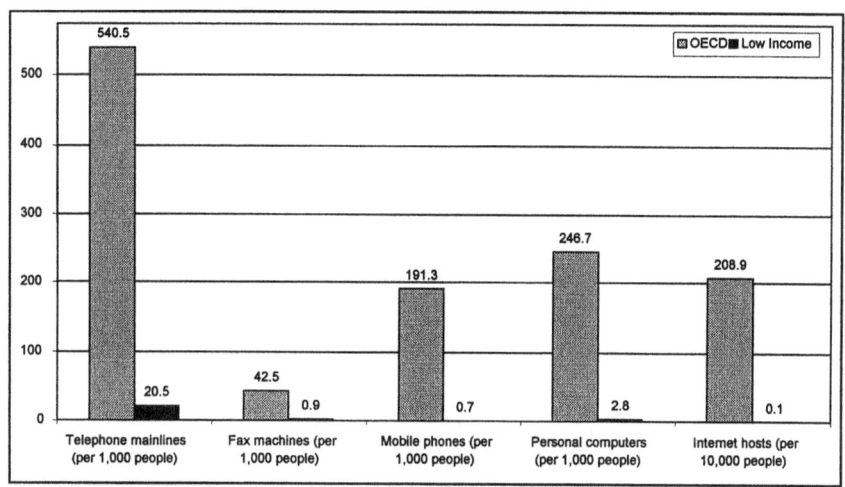

Figure 2-7: Inequalities in ICT infrastructure between OECD and low-income countries (Data source: WORLD BANK 1999)

On the sub-national level, however, this picture becomes ambiguous. Before this level will be discussed, we need to mention why low-income countries are said to be able to gain even more from ICT implementation and use than rich countries and – as a consequence – be able to leapfrog conventional phases of infrastructural and economic development (BEDI 1999; THE ECONOMIST 2000; G8 2000).

First, low-income countries start off with a low level of capital per worker, which means that there is therefore a certain "scope to grow rapidly by buying rich countries' technology and copying their productions methods" (THE ECONOMIST 2000, p. 41). As there is no need to reinvent the wheel, the late adopting poorer countries could, provided they are open to the technological advances from the industrialised world, in this sense grow faster than developed economies, even if they start off from a lower technological base.

Second, ICTs are infrastructural means that spread much faster than previous technologies such as railway or electricity. Advances in solar, satellite and wireless technologies do make it possible for low-income countries to simply skip intermediate technology stages. For instance, the implementation of telecommunication services does not necessarily require copper wire-based infrastructure. The infrastructure is increasingly being realised using digital wireless technologies (THE ECONOMIST 2000; WORLD BANK 1998). This leads to the somewhat

47

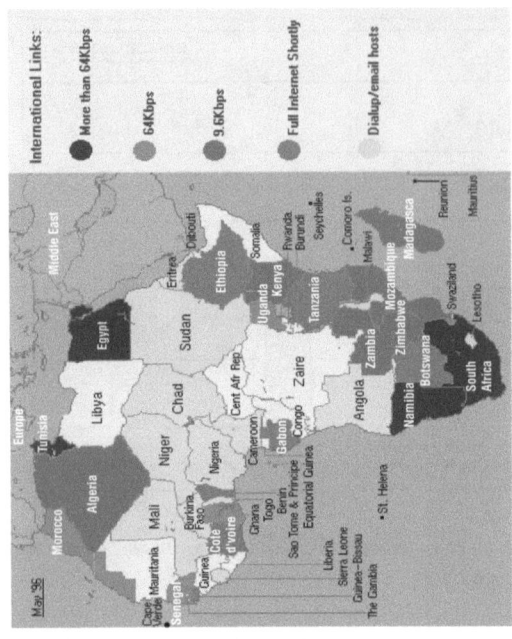

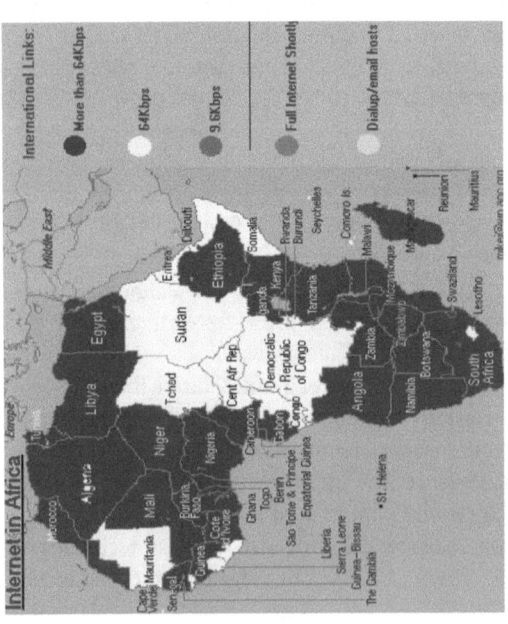

Figure 2-8: *Internet connectivity in Africa in the years 2000 (left) and 1996 (right)*
(Source: JENSEN 2000)

surprising situation that many low-income countries account for a higher rate of digital main lines than rich countries. Seen on average, they only slightly lag behind the OECD countries, and in fact could catch up during the last years (Figure 2-9).

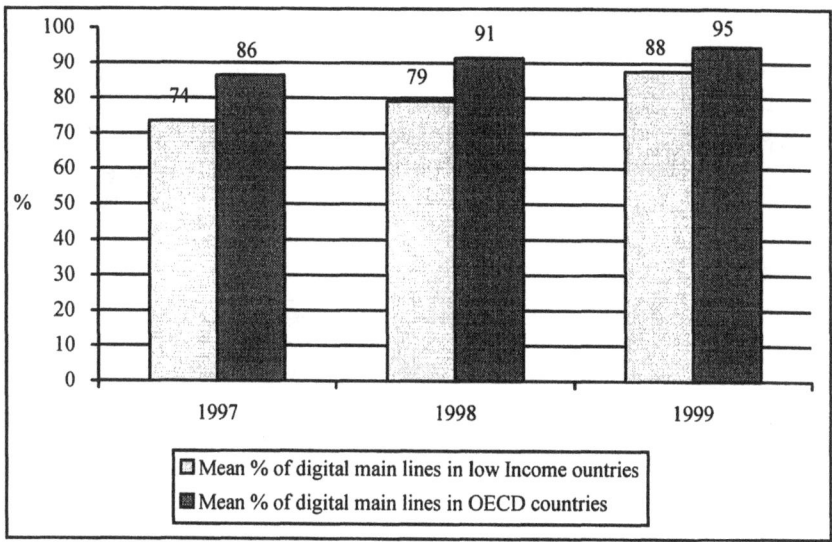

Figure 2-9: Mean percentage of digital main lines in OECD and low-income countries (Data source: ITU 2000)

Another example of developing countries catching up is seen in the mobile phone market. WELLENIUS ET AL. (2000) and SAUNDERS ET AL. (1994) point out that – compared to OECD countries – the developing and transition economies accounted for 40% of all phones in 1999 but only 11% in 1988 and 7% in 1981 respectively. The authors conclude that this narrowing of the gap is mainly resulting from the addition of millions of mobile customers in countries that are not members of the OECD. For the group of the low-income countries, Figure 2-10 shows this in a less dramatic but still valid manner.

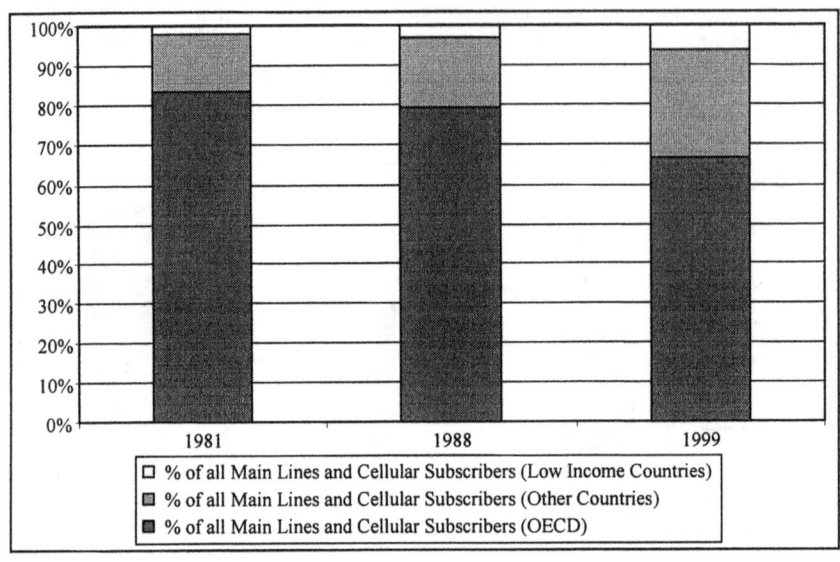

Figure 2-10: Shares of main fixed lines and mobile cellular subscribers for OECD, low-income and other countries (Data source: ITU 2000)

2.4.2 The Rural Challenge

At second glance, the analysis of the diffusion of the *leapfrogging technologies* shows that they are limited to the metropolitan areas and larger cities of low-income countries. In remote and rural areas, the diffusion of ICTs is hampered by basic infrastructural shortcomings such as an underdeveloped and unreliable electricity system but also by the lack of access to basic telecommunication services (cf. Figure 2-11; FAO 1998; G8 2000; ITU 1998a; SEIBEL ET AL. 1999).

Bearing those circumstances in mind, one can ht recognise that relatively cheap access to ICTs in general and all kinds of Internet services in particular are restricted to multinational firms, agents in the development community and major players in the public sector, such as ministries and universities. Currently they represent the nodes of connectivity of low-income countries to the emerging global information infrastructure. And even if ICTs in these areas increase productivity and contribute to economic development in general, the benefits may not trickle down to the parts of the population with low incomes and especially not to rural areas:

"It is generally agreed that [...] ICT access unevenly favors urban and wealthy residents" (TELECOMMONS DEVELOPMENT GROUP 2000, p. 3).

Behind the latter notion the "rural penalty" gets yet another dimension (HUDSON 1992). The lack of basic infrastructure and social services for rural areas is often due to physical and financial difficulties and the low returns on investment that can be expected. This leads to the well-known situation that residents in developing countries' rural areas lack access to basic needs such as water, food, education, health care, sanitation and security. Additionally, they lack information and knowledge, which is an indispensable ingredient in development at any level. On the one hand, this deficiency becomes increasingly important as even regions become more integrated into national and international markets. On the other hand, intimate information systems within the traditional society are no longer available whilst modern institutions for reducing information deficiencies are not yet established or do not work properly (GEERTZ 1978; WORLD BANK 1998).

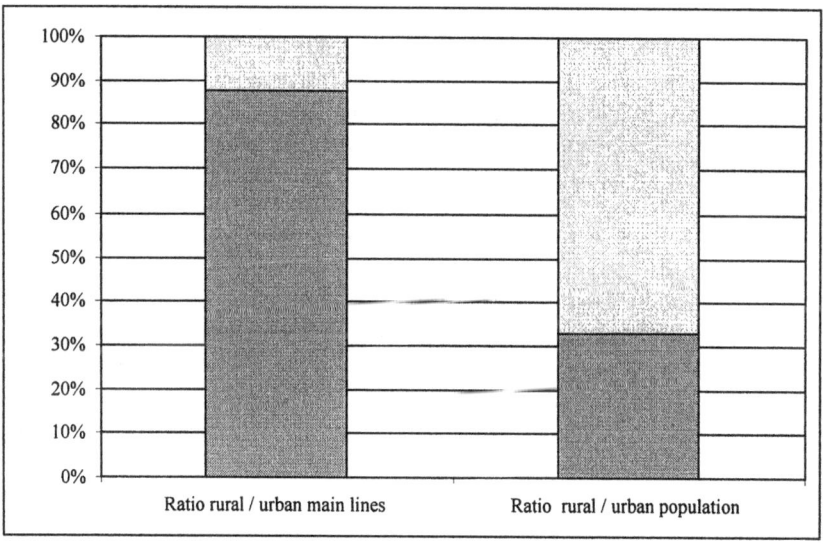

Figure 2-11: Urban versus rural telephone lines in low-income countries (Data source: ITU 2000)

In view of these circumstances, the ability of ICTs to bridge space and time (cf. Section 2.1) is often seen as particularly beneficial in a situation where information is usually communicated face-to-face or by sending messengers: the introduction of even the most basic form of ICTs, i.e. voice telephony, into such environments can help people to reduce time and costs spent on communications (cf. Sections 2.3.2.2, 2.3.3.2). Moreover, ICTs can significantly increase the overall amount of information available and thus simplify decision-making processes in a situation in which "information is poor, scarce, maldistributed, inefficiently communicated and intensely valued" (GEERTZ 1978, p. 29).

While the potential impact of ICTs on rural development is comprehensive, there are several concerns related to the introduction of ICTs.

From the perspective of the determinants and dimensions of diffusion introduced earlier (cf. Section 2.2), we need to recognise the following issues:

In most developing countries, the formation and implementation of policies related to an equitable ICT supply in rural areas is still very rudimentary and calls for an integrated set of laws, regulations and guidelines. Apart from these policy considerations, the accessibility of the technologies raises questions about the successful implementation of ICTs in rural areas (BRAGA ET AL. 2000; SAUNDERS ET AL. 1994; TELECOMMONS DEVELOPMENT GROUP 2000).

In this respect, accessibility to ICTs is a complex problem of infrastructure, cost, literacy, gender, etc. From the perspective of service operators, provision of ICTs to rural areas is costly and the expected revenue stream is low, which often results in market failures. In addition to the equipment for ICT applications, electricity and telecom networks of reliable quality at low prices are still a prerequisite for new ICTs such as the Internet. It is also surprising that the latter was often the focus of policy agendas (cf. ITU 1998; UNDP 1998; WORLD BANK 1998) whereas the basic infrastructural problems were rarely addressed. Moreover, computer technology is seen as the greatest driver for changes on the level of telecommunication services (GILLIS, MCLELLAN 1998). This, however, means taking quick and cheap provision of access to telecommunication services for granted, which by no means could be seen as an adequate approach in light of the mentioned infrastructural facts (FAO 1998).

Access to and use of ICTs is not only a question of service availability, but also a question of occupation, income, and education. The latter aspects in particular are major constraints in most rural areas and especially limit the number of people who can afford to use the service, provided they are physically present at all.

It is, however, not possible to generalise in this context. There is an increasing awareness of the fact that a distinction has to be made between those ICTs whose user interface is a costly and difficult-to-use computer and those that could be used and afforded by almost anybody but are still lacking in most rural areas, i.e. basic telephone services:

[...] despite high demand for ICT services and a growing telecommunications market, rural access to basic ICT services, such as voice telephony, remains a major concern" (TELECOMMONS DEVELOPMENT GROUP 2000, p. 3)

It is the major task of this work to elaborate on this concern and point out the vast difference between the potential seen in ICT implementation and the sheer lack of a telephone infrastructure in place. This infrastructure is based on tele-communication networks and allows the transfer of verbal and coded information by either analogue or digital means without the precondition of being able to op-erate a computer. The networks are the first step to getting rural areas connected and are certainly the precondition for any other more sophisticated ICT applica-tion. We should also bear in mind that universal service, "usually defined as a telephone in every home" (ITU 1998a, p. 10), is not seen as an adequate solution towards the supply of telecommunication services in low-income countries in general and their rural areas in particular: under poor countries' circumstances, one has to focus on providing the majority of the population with services. The fact that the development of nation-wide service availability at prices that can be afforded by the average citizen cannot be achieved in the short-term led to the ar-ticulation of transitional goals. At the centre of this articulation is the concept of universal access: "The concept is that a telephone should be within a reasonable distance for everyone. The distance depends upon the coverage of the telephone network [...] the density of the population and the spread of habitations in the ur-ban or rural environment. This diversity has been reflected in a range of innova-tive policies and platforms: from the use of public telephones to entrepreneurial teleshops to community telecentres" (ITU 1998a, p. 10).

This chapter aimed at showing the link between the uniqueness of ICTs, their diffusion and economic benefits. We also showed how this economic potential could be assessed conceptually and what empirical evidence exists in this respect.

Finally, the need to narrow down the scope of the research became apparent whilst considering the infrastructural, political, and socio-economic problems that are apparent in many low-income countries in general and their rural areas in particular.

Against this background and with the focus on basic telecommunication ser-vices, we will introduce the existing trends in telecommunication provision in low-income countries and their impact on infrastructural developments with par-ticular reference to Ghana. This will form the basis for the intended empirical as-sessment of the supply of service infrastructure and its use at the community and the household level.

3 Telecommunications in Sub-Saharan Africa: Institutional Trends and Infrastructural Developments

From the previous chapter we know that there is a growing interest in strengthening the telecommunication infrastructure of low-income countries. This is only partly due to the realization that reasonably well-managed telecommunication entities can generate large financial surpluses in local currency. Moreover, telecommunication networks are seen as a driving force towards macroeconomic growth through direct and indirect effects.

It also became clear that low-income countries are lagging behind all other countries in terms of their telecommunication infrastructure. This especially applies to Sub-Saharan Africa: "Densities [telephone main lines per 100 inhabitants] for the region are among the lowest in the world, and most rural areas have no access to services at all" (SAUNDERS ET AL. 1994, p. 320). It is therefore important to remember that this work primarily sets out to ask who actually has access to ICTs in rural areas. Starting from the supply side and assuming that the prevailing conditions of incomplete markets do require additional factors that explain the diffusion of services, a more comprehensive overall view of telecommunication sector developments will be provided.

This chapter will therefore show how telecommunication sectors in Sub-Saharan Africa have been modified over the last couple of years. Reforms, namely private participation, competition and regulation, require governmental support. It is therefore crucial to first discuss the trends by which a growing number of low-income countries are modifying their telecommunication policy. Regarding the telecommunication sector, the institutional and organisational patterns do not differ significantly from strategies applied in developed countries. Moreover, most reform "programmes tend to be modelled on continental European precedents" (MUSTAFA ET AL. 1997, p. vii). The nevertheless existing particularities of those processes in low-income countries will be related to general trends observable in Sub-Saharan Africa. This will be done by analysing the transformation processes on an organisational, institutional, and infrastructural level.

The points mentioned above will comprehensively introduce the analysis of telecommunication sector reform in Ghana, which is strongly committed to the liberalisation of trade and markets and is, together with a few other countries, widely regarded as a best practice case (ITU 1998; WORLD BANK 1998).

The analysis of the specific telecommunication infrastructure developments on a national and sub-national level in Ghana will eventually form the basis for demonstrating the extent to which institutional and infrastructural modifications manifest themselves on the local level.

3.1 Trends in Telecommunication Sector Reform

Up to the end of the 80s, "telecommunications used to be regarded as a natural monopoly and a relatively straightforward public utility. Economies of scale, political and military sensitivities, and large externalities made telecommunications a typical public service" (SAUNDERS ET AL. 1994, p. 305). Consequently, the services were mostly[8] provided in a monopolistic manner by government departments or state-owned enterprises. In such a way, industrialised countries ran profitably operating infrastructures and succeeded in providing universal access to telephones. Developing countries, however, started from a different level. Their telecommunication services were initially run by foreign private companies and colonial government agencies. They provided an infrastructure that was primarily designed to meet the needs of colonial administration rather than the population. During the 60s and 70s, most telecommunication providers were taken over by the public sector of the then independent national governments. These monopolies fell short of meeting the demand for telephone services which resulted in long waiting lists, congested call traffic, poor service reliability, and limited territorial coverage. This was due not only to the scarcity of capital but also to the inefficient management of existing resources and returns as well as a limited motivation to meet the existing demand (MUSTAFA ET AL. 1997; SAUNDERS ET AL. 1994).

Albeit a poor market performance, MUSTAFA ET AL. (1997, p. viii) point out "that there is evidence of substantial ability and willingness by customers to pay for services which were [simply] not available". In this respect it seems encouraging for the authors that revenues per line are high compared to world standards and they conclude: "the telecommunication sectors of SSA countries have the potential to be profitable" (MUSTAFA ET AL.1997, p. viii).

These positive perspectives and the increasing discussion on ICTs as a catalyst for a more efficient economic performance and increased global integration (cf. Chapter 2) made sectorial reform a widely discussed issue. In the North this was mainly due to the fact that the old monopolistic structures had been over-

---

8   Major exceptions in this respect were Canada, Finland, and the USA (cf. SAUNDERS ET AL. 1994).

come and could not meet the challenges that were imposed by the huge demand for new services. In the South, the emergence of the *leapfrogging technologies* in line with the urgent need for a higher penetration rate of basic services were particularly challenging. For both hemispheres, telecommunication sector policies were designed along rather similar patterns. Although some low-income countries – especially in Sub-Saharan Africa – still hesitate to adapt these kinds of patterns (cf. 3.1.1), some governments increasingly overcame their reluctance and joined ICT-related programs and initiatives set up by international organisations such as UN-ECA, UNDP, ITU, and USAID. Substantial parts of these programs and initiatives address concerns that governments are not able to face the financial and managerial challenges in order to implement the sectorial modifications required. More directly, they help to attract private investment and to accelerate the expansion of the infrastructure and services (ITU 1998; JENSEN 1998).

Some Sub-Saharan African countries that got involved in such programs, namely Côte d'Ivoire, Ghana, Mauritius, Sénégal, and South Africa eventually embarked on international agreements and restructured "their telecommunication markets under the framework of the World Trade Organization (WTO) agreement on basic telecommunication services and the General Agreement on Trade in Services (GATS)" (ITU 1998, p. 8).

From a global perspective and with regard to telecommunications, this agreement was introduced in order to meet the need for a new international system of governance for telecommunication generated by economic globalisation and technological convergence (TARJANNE 1999). It was signed in early 1997 by countries that account for more than 90% of the world's telecommunication service revenues. For these countries the most important principles that should be introduced under the umbrella of GATS and in order to promote the aforementioned acceleration of infrastructure and service provision are as follows (ITU 1998; PETRAZZINI, KELLY 1998; SAUNDERS ET AL. 1994; TARJANNE 1999):

- The separation of the former monopolist, the national post and telecommunications operator (PTO), into telecom and postal service providers.
- Fostering privatisation of the former incumbent telecommunication service provider and promoting of foreign investment.
- The liberalisation of the markets, i.e. the introduction of competition within the sector.
- The establishment of a legal framework that efficiently regulates the developments in the sector.

The respective governments – under the assistance of the ITU – progressively follow those structural means which will now be introduced. Although there are differences between the individual countries, the elements mentioned below appear to be an institutional *best practice* (ITU 1998, 1999). It is, however, important to note that the distinction made between the different measures of telecommunication sector restructuring does not reflect the reality. Furthermore, the processes are interlinked and not as separable as it may appear from the following sections.

3.1.1 Separation, Privatisation, and Liberalisation

Separation

The separation of the telecommunications operator from postal services and the transfer of both entities to corporate bodies is the first step towards sectorial reform. As a consequence, the telecommunications operator has to be reorganised following commercial principles:

"There is widespread agreement that regardless who owns them, telecommunication operating entities perform best when they are run as profit-driven businesses"(SAUNDERS ET AL. 1994, p. 310).

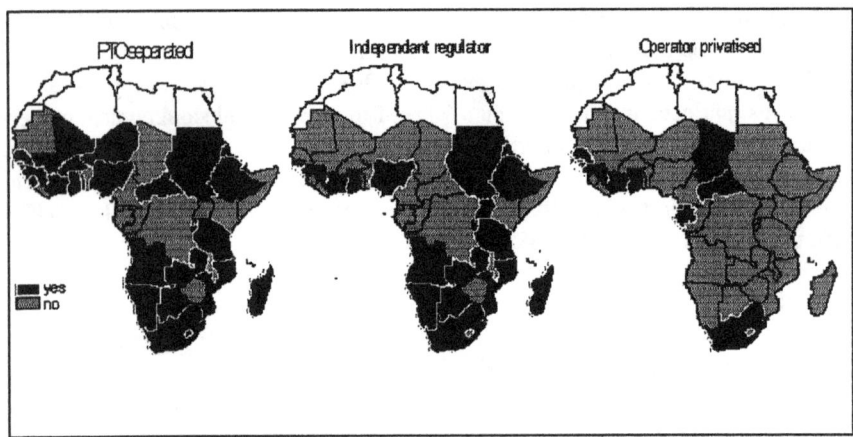

Figure 3-1: Regulatory reform in Sub-Saharan Africa (Data source: BMI-T 1998; ITU 1998)

According to the ITU (1998), the transformation of the incumbent entity is always achieved along the lines of a continuum which begins with the reorganisation of the government departments into state enterprises. In a further move, the latter are transformed into state-owned companies to finally enable private capital to partly or fully own the entity. Along with those external processes, internal restructuring is usually taking place, eventually leading to the reorganisation of the enterprise into cost and profit centres. Concerning the development and training of the staff members, emphasis is placed on customer service, cost awareness, financial cautiousness and their technical and managerial performance.

As shown in Figure 3-1, quite a number of Sub-Saharan African countries have separated their post and telecom operations, given it a corporate structure, and privatised the entity (for details cf. Annex 3-1)

Privatisation

It has already been pointed out that separation and corporatisation aim to attract financial investment into the telecommunication service provider.

The most common strategy to privatise national telecommunication operators in low-income countries is to offer a share of the company to a strategic foreign investor as an equity partner. This is often regarded as the only way of generating the required investments needed to develop telecoms in view of scarce domestic resources in terms of capital and hard currency (ONE WORLD 1997). This strategy also allows the investor to bring in managerial and technical skills and expertise on which the operator can build: "Where technology transfer is the main objective, a partnership scheme is more likely to be chosen whereas a government hoping to raise as much cash as possible will probably opt for a public offering" (ITU 1997, p. 47).

Most strategic investors are telecommunication operators from industrial countries. After positioning themselves in their own changing domestic and regional markets, many of them pursue new business opportunities in developing countries (SAUNDERS ET AL. 1994).

Whatever formal implementation the privatisation process has, the positive effects expected from this process, are as follows (ITU 1998, 1997; TARJANNE 1999):
- Enhanced efficiency in management, production, and the provision of services.
- Financial benefits for governments, so that it is potentially able to invest into other sections of the public sector or to reduce foreign or internal debt.
- Reduced disguised and open subsidies.
- The expansion of telecommunication infrastructure to underserved areas and increasing network quality.
- The subsequent reduction of tariffs.

Despite these expectations, some countries feel reluctant to sell the operator because for them, "government control of public services is linked with independence and nation building" (MUSTAFA ET AL. 1997, p. vii). From a more specific viewpoint, telecommunication services are still seen as a resource of strategic importance (ITU 1997; ONE WORLD 1998). Along the lines of the infringement of national sovereignty and the loss of control over basic communication structures, the lack of financial profits steadily flowing in from the operator can be named as reasons for this reluctance. But even in countries that are committed to the privatisation and liberalisation of their telecommunication sector, state-owned companies are rarely sold entirely: governments usually retain some degree of influence on the company either by keeping the majority of shares or by creating a so-called *golden share* that allows them to veto at least major decisions (BMI-T 1998; ITU 1997; ONE WORLD 1998).

With regard to Sub-Saharan Africa, these factors seem to be of minor importance if one considers international operators' fears linked to investments in that particular region. Referring to an analysis by the World Bank, GRANT (1998) points out that the operators see the Sub-Saharan region as generally difficult to work in. The main reasons for this are political instability and problems with corruption.

It is not the subject of this dissertation to judge which of the countries in Sub-Sahara Africa the mentioned problems apply to. It might be worthwhile, however, to point out that although most African countries are taking measures to increase private sector participation, investment inflows in the field of telecommunications actually fell from US$ 5 to US$ 4.7 billion in the 1994/1995 period (ONE WORLD 1997). There are, however, big differences between individual countries as far as their commitments towards telecommunication sector reform is concerned (cf. Annex 3-1).

Liberalisation

As already indicated, the opening of the sector towards new market players has to be perceived as an integral part of sector reform. There are two mainstream strategies of liberalising the telecommunication markets, i.e. complementary and competitive entry policies. Both aim at attracting fresh capital and new management resources for the telecommunications sector; mainly through mechanisms that are due to licence obligations or that create competition amongst the service providers.

The complementary approach's "aim is to encourage investment in infrastructure, especially local networks and new services. It is therefore an effective instrument wherever telecommunication service supply is underdeveloped"

(MUSTAFA ET AL. 1997, p. ix). As far as options of complementary entry strategies are concerned, one can distinguish between those aiming at a divestiture of the market, be it sectorial or regional and franchising agreements. The latter allow agents to offer services on the basis of the infrastructure owned by the former monopolist (BMI-T 1998).

3.1.2 The Regulatory Framework

Definition and Nature of Regulatory Reform

So far, we have considered regulatory reform as comprising sector policies and market regulation. The New Institutional Economics "characterises regulation as a form of contract between government, customers, regulated companies designed to facilitate a service agreement between them" (MUSTAFA ET AL. 1997, p. 56; cf. WILLIAMSON 1979).

These views are difficult to operationalise within the context of this work. Following the more pragmatic approach of MUSTAFA ET AL. (1997), we distinguish between market trends as described above, the corresponding implementation strategies (cf. 3.1.1), and regulatory reform. The latter includes changes in rules, procedures and administrative arrangements to which the market players are subject.

Generally speaking, regulatory reform is introduced by governments to ensure that the desired sector developments can be implemented and the shift from the state monopoly to competition can be executed. More specifically, regulation is considered to be a major and inevitable political task that should be pursued by an independent government authority. The latter facilitates and controls access of profit-motivated or shareholder-driven telecommunication companies to compete with the former monopolist. Sector regulation also enables governments to link an investor's market entry to obligations that have a specific development focus, e.g. providing services to poorer people and to rural areas (ONE WORLD 1998).

Against this background the tasks and contents as well as practical aspects of the implementation of regulatory reform will be pointed out[9] in the following section.

---

9   Although an important element of any country's telecommunications policy, technical standards for equipment, compatibility of network interfaces as well as issues of transmission frequencies are not included in this discussion. This is mainly because of their limited relevance for the empirical part of this analysis.

3.1.2.1 Tasks and Contents of Regulation

Tasks and contents of the regulation process obviously differ with the degree of regulatory reform in place, i.e. government ownership, corporatisation and privatisation of the operator, as well as the opening of the market to new entrants and competition.

Where monopolists provide telecommunication services they can take the organisational form of either a government department or a public corporation. A third form is an intermediate position "in which the operating entity has the status of a government department but can act as a commercial agency" (MUSTAFA ET AL. 1997, p. 56).

The regulatory rules that are important in this respect are mainly business elements such as the market scope of the monopolist, the external financial resources accessible and the form of management of the entity (ITU 1998).

The possibility of running the monopolist as a public corporation was often considered as the first step of reform (cf. 3.1.1). This regulatory form contains statutes which limit the influence of government officials on the corporation to an *arm's length relationship*. However, sector ministries are or were often not willing to loosen control and hinder the development of a more efficient and independent operator (ITU 1998).

The next step of sector reform involves the introduction of commercial structures (cf. 3.1.1). The former incumbent is now operating under company law: the responsibilities of its management "can be discharged through the appointment of board members and through the contracts of senior managers" (MUSTAFA ET AL. 1997, p. 58). Additionally, market strategies that will limit the monopoly status of the operator are set up. Through the commitment to such strategies, governments provide the enterprise with operational security and the opportunity to incorporate future developments, e.g. to prepare for competition (ITU 1999).

Box 1: Joint Ventures

In most instances, new telecom-munication service operators are established as joint ventures between strategic foreign investors and local partners. This especially applies to the cellular phone market: practically every cellular network in Africa is backed by strategic foreign investors. However, fixed line operators can also operate in such a form, which is said to be especially favourable because it enhances the transfer of capital, technology, know-how, and management expertise.

At this stage, the privatisation of the company is often due to be introduced. In most developing countries that are committed to telecommunication sector reform, this happens through a partial sale of government shares to strategic investors leading to a joint venture agreement (cf. Box 1). The company is then subject to the conditions, e.g. licensing or concession contracts that were negotiated with the government. It will be particularly cautious with respect to future sector policies, i.e. tariff rules, degree of liberalisation, liabilities, exit options, and commitments towards network expansion (ITU 1999; SAUNDERS ET AL. 1994).

The latter aspect is of specific importance in light of the typical sector objectives that considered telecommunications as a social good to be made available to the people at affordable prices. This ideal of universality has prompted state-run operators to cross-subsidize less or not at all profitable service provision, i.e. in poor neighbourhoods or remote rural areas, using profits from more lucrative telecommunication services. With privatisation, the issue of cross-subsidy will be challenged as investors might just want to "cream skim" the most profitable areas of the market rather than broadening access to undeserved areas (ONE WORLD 1997). A potential withdrawal from those customers might eventually lead to a further decline in access rather than its improvement.

Such undesired developments are taken into account by gradually introducing competition. In most cases this happens unevenly across market segments.

One could first consider a complementary introduction of competition.

This happens, for instance, if monopolies are divided up by regions which enables governments to separate profitable from less profitable markets in order to be able to offer regional stakes more successfully. Governments could also award licenses to operators that provide specialised networks to meet the needs of communication-intensive segments of the economy, i.e. the banking, tourism, or the mining sector. Moreover, independent public telephone companies could be awarded with licences that are linked to areas that would otherwise remain unserved.

Finally, extensions of the public telephone networks are licensed in order to speed up network expansion in unserved areas (ITU 1999; SAUNDERS ET AL. 1994).

The option of competitive entry allows one or more entrants to provide service in competition with the monopolist. This strategy, in particular, increases commercial pressure on the former incumbent. Competition as such not only fosters investment but through its rivalry effect is likely to spur new and established operators "to focus attention on customers, improve service, accelerate network expansion, reduce costs and lower prices" (SAUNDERS ET AL. 1994, p. 311). Apart from enhancing existing services, it facilitates the introduction of new ones and speeds up their diffusion[10]. With regard to basic telecommunication services, the scope of this subject can be limited to the licensing of a second national op-

---

10 The latter aspect could widely be observed in the mobile cellular and value added, i.e. Internet service markets.

erator (SNO). This is the most common way to initiate competition in basic tele-communication services. As a large-scale coverage of all market segments is difficult to achieve for the entrant, competition is often introduced in long distance service provision and business customer markets. The high amount of traffic and revenues in these segments reduce the importance of economies of scale. Over time, SNOs should, however, be able to serve the market as a whole and, thus, also compete for small businesses (ITU 1999; MUSTAFA ET AL. 1997).

3.1.2.2 Issues of Licensing and Interconnection

As the prospect of foreign investment is a major incentive for market liberalization in sub-Saharan Africa, regulation should always be based on a reliable legal framework. Licences and concession contracts are the most common tools in this respect. Whether permanent or temporal, nationwide, or regionally restricted, licences build an important planning basis for market players and investors and are also tools that enable governments to tie market access to certain conditions and to charge licence fees that can be used to finance the regulator and to gain resources for network expansion (ITU 1999).

To particularly promote this expansion, regulatory agreements often tie market access or guarantees to certain duties that the operators need to fulfil. One application of licences, particularly important for rural areas in low-income countries, is to promote the universal access idea. In practice this means that they are used as tools to oblige service providers with duties to implement a certain number of publicly accessible telecommunication facilities in poor and/or rural areas within a specific period of time. This approach is often regarded as "one of the few areas where sector-specific regulation may be required indefinitely, even when competition has spread across market boundaries" (ITU 1999, p. 11). For typical terms and provisions of a licence agreement cf. Figure 3-2.

Furthermore, the modalities of interconnection between the networks of different service providers are important regulatory tasks (cf. Figure 3-2). Focusing on non-technical aspects in a multi-operator environment means addressing the divergence of the commercial interests of different market players. This usually requires regulatory intervention in order "to determine revenue shares in accordance with overall policy aims" (MUSTAFA ET AL. 1997, p. 70). It is therefore crucial to set out interconnection arrangements and charges in the basic regulatory documents, e.g. in a country's telecommunication development plan or within the licence agreements.

The definition of interconnection charges follows the principles of tariff policies. Accordingly, charges should be set up equal to the incremental cost of expanding output. Logically, the basis for determining interconnection charges is the additional costs borne by a network operator that can be directly attributed to establishing and operating the interconnection. Under a regime of financial short-comings and overall economic instability, it might not be possible to identify these costs satisfactorily. A solution proposed by MUSTAFA ET AL. (1997) to tackle this problem would be to set up tariff-based interconnection charges, which is often done for cellular networks. This approach also has the advantage that tariffs are a transparent and clear point of reference and administrative costs are reduced due to the fact that tariffs are already subject to regulatory controls.

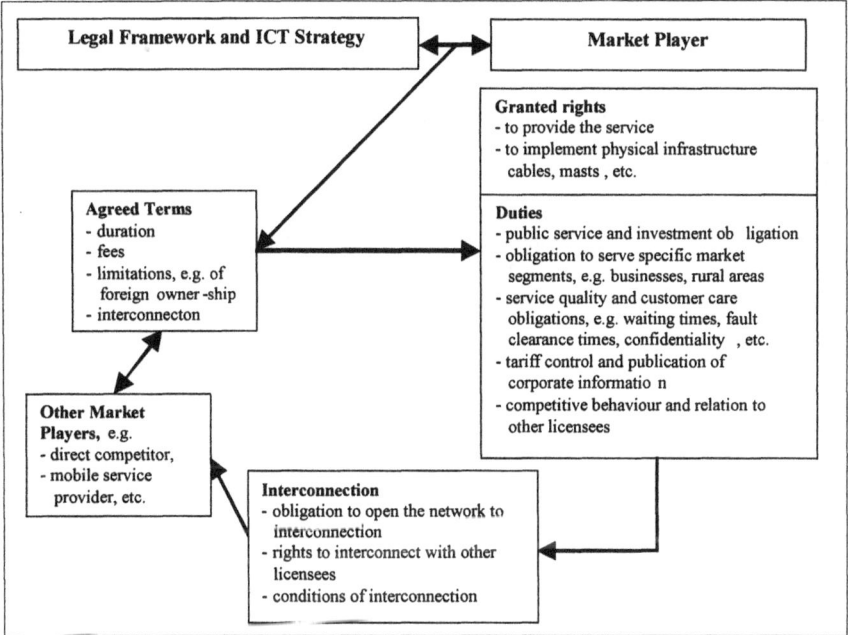

Figure 3-2: Typical terms and provisions of an operator's licence (cf. MUSTAFA ET AL. 1997)

All in all, it should have become clear that market regulation proves an am-bivalent task: On the one hand, the former monopolist's market power has to be limited by a legal framework in order to facilitate market access to competitors; on the other hand, some freedom should be left to the monopolist to enable the corporation to adjust to the new market conditions. A crucial step towards the so-lution of this somewhat contradictory situation is the establishment of regulatory mechanisms.

3.1.2.3 Establishing Regulatory Mechanisms

The establishment of such mechanisms is mostly carried out by forming a regulatory authority, which will effectively formulate and regulate the sector developments by applying the tools described above (cf. 3.1.2.1, 3.1.2.2). From an organisational perspective, these authorities are quasi-separate bodies with limited powers. Often, the ministry responsible retains the authority to issue directives to the regulator to control the funding and the board of the agency. The degree of the regulator's autonomy depends on the manner in which it was established, whether by a comprehensive law or by a ministerial decree (ITU 1999; SCHERER 1994). It is argued, that only complete independence of the regulatory authority can provide sufficient assurance to market players so that ad hoc interventions by governments are excluded. In their WORLD BANK discussion paper, MUSTAFA ET AL. (1997) indicate that not a single effectively independent regulator could be established in developing countries that have undergone privatisation up to 1997. Furthermore, they state that "private investors have not required regulatory independence as a condition of their participation in the sector" (MUSTAFA ET AL. 1997, p. 63). They rather ask for assurances that governments will not change rules after they have invested in the market: instead of guaranteed independence from the regulator, the assurance that the investor's "interests will not be adversely affected by changes in policies is best embedded in licences or concession contracts" (MUSTAFA ET AL. 1997, p. 63). As cases from industrialised countries indicate, the main reason for an effectively independent status of the regulator "seems, thus, to be that, in a liberalised or privatised sector, quasi-legal judgement of commercial behaviour and settling commercial disputes is required" (MUSTAFA ET AL. 1997, p. 63). As this task might be fulfilled by judicial bodies that settle commercial and competition law disputes, the sense of establishing a resource intensive separate body is open to discussion. However, the ITU points out that especially for Sub-Saharan African countries these kinds of mechanisms also indicate the increasing desire to attract foreign investment into the telecommunication sector (ITU 1998).

3.1.2.4 Limits and Difficulties

Problems concerning the regulatory process that are particularly evident in low-income countries are discussed in numerous publications (cf. ITU 1998; SCHERER 1994). They mainly focus on the following issues.

Despite the fact that regulatory mechanisms and institutions have been established in many low-income countries, the proper establishment of institutions proves to be a difficult and time-consuming task. Governments, for instance, often hesitate to revise the traditional opinion of telecommunications being a natural monopoly due to the sector's outstanding social, economic and strategic importance (cf. Section 3.1.1). As a result, telecommunication sector reform is a highly political issue and the informal influence of governments on licence awarding processes or investment priorities might, in many countries, be counterproductive to the overall goals set.

In the case of numerous Latin American countries[11], however, privatisation has taken place rather successfully despite the absence of a functioning telecommunications regulatory system, but at the cost of developing a competitive market:

"The largest privatised companies are operating with little or no competition and in a regulatory vacuum in which critical regulatory responsibilities regarding licensing, pricing, technical and accounting standards [...] are not being properly discharged. In a market dominated by one operator and lacking effective and proactive regulation, competitors are unlikely to emerge" (SAUNDERS ET AL. 1994, p. 323).

Furthermore, new regulatory agreements and institutions are often set up ahead of real operational progress and a distinct policy strategy, hoping that they will set signals for potential market entrants and investors from outside the country. This leads to the situation where "several countries [in Sub-Saharan Africa] have undertaken regulatory reform in a relative policy vacuum and in advance of substantive restructuring of operations; in consequence, the reforms seem overcomplicated and are likely to prove ineffective in promoting improvement in sector performance"(MUSTAFA ET AL. 1997, p. xi).

Additional scepticism concerns the fact that a newly established regulatory body lacks political authority and more powerful interests may initially determine its policy direction. Accordingly, the early establishment of regulatory institutions creates a situation in which the tasks these regulatory institutions are responsible for are not yet clearly specified.

Despite these problems, there is a clear need for an elaborate and balanced system of regulation. However, it is important to consider that until a market structure has developed, any rules set up by legislators or regulators are often pointless. Worse, an excess of regulatory rules can kill off the interest essential for getting a market off the ground. Therefore, a consensus drawn from the authors mentioned could be that rules in emerging telecommunication markets should only be created as and when they are required.

---

11   SAUNDERS ET AL. 1994 mention Argentina, Mexico, Venezuela and Chile.

3.2 Ghana's Telecommunications Sector Policy and Infrastructure

The extent to which the aforementioned trends in global and regional telecommunications reform are applicable to Ghana will be the focus of this section. These mechanisms will be linked to the country's progress in infrastructural development on a national and sub-national level.

This, requires a brief look at the country's overall commitment towards open markets and liberalised trade.

J.J. Rawlings – for the second time in 1982 – took over political power in the Republic of Ghana when the country was undergoing its deepest economic crisis since it gained independence in 1957. These problems led the government, towards the end of the same year, to start an economic restructuring process that mainly followed the patterns of structural adjustment as proposed by the IMF and the World Bank. Loans provided by the IMF were the necessary precondition to regain credibility of the donor community and to acquire the monetary resources necessary for the recovery of the economy (SCHMIDT-KALLERT 1994). Since then, the country has carried out a couple of "the most thorough structural adjustment programs in Africa" (CHIBBER, FISCHER 1991, p. 7) and also been committed to the GATS. Important tools in this respect are (IMF 2000; SAPRIN 2000):
- the privatisation of government-owned companies,
- the reduction of public sector expenditures,
- the fostering of deregulation policies,
- the enhancement of macro-economic stabilisation through adequate monetary and fiscal policies,
- the liberalisation of imports and the promotion of exports.

These elements can also be found in the latest Structural Adjustment Facility Policy Framework that was elaborated in 1999 for the period until 2001 (IMF 2000) and certainly paved the way to telecommunication sector reform.

3.2.1 The State of the Ghanaian Telecommunications Sector

One of the first measures of the government's commitment towards ICT development was to restructure the Ministry of Transport to form the Ministry of Transport and Communications. With the assistance of various international organisations, such as the ITU, the ministry then formulated a national strategy for broad information and communication sector reform. The strategic cornerstones of this reform were then fixed within the Telecommunication Policy for an Accelerated Development Program (ADP) that was launched in the year 1994 (MINISTRY OF TRANSPORT AND COMMUNICATIONS 2000). The aims and measures of this sector policy program will now be discussed, particularly pointing out the effects seen so far from both an organisational and infrastructural perspective.

3.2.1.1 Sector Policy: Aims and Measures

The sector strategy aimed to target the following issues (FREMPONG, ATUBRA 2001):

First, a tariff policy was suggested that allows the market players to recover the full cost of providing the service.

Second, it was fixed in the ADP that customers should be able to access services at competitive and affordable prices. This included the aim to foster public accessibility in urban and rural areas through the implementation of pay phone facilities. In the long run, public access in Ghana should also be improved by providing universal access through the installation of "pay-phone facilities to every village of a minimum of 250 inhabitants" (IICD 1999, p. 5). Besides this goal of universality, the establishment of high-quality communication services for the business community is in line with Ghana's long-term economic policy objective of increasing its competitive advantage on an international level and of becoming the *Gateway to West Africa* (FREMPONG, ATUBRA 2001).

The most important means of achieving these ambitious goals was the establishment of Ghana Telecom in June 1995. Ghana Telecom took over the telecommunication division of the Ghanaian PTO, which was turned into a state-owned corporation as early as 1974. This corporatisation did not, however, have positive effects on either the profitability of the company or its supply of services (cf. Figure 3-3)(GHANA TELECOM 2000a, also cf. 3.1).

To attract financial investment into the newly established, but heavily indebted, telecommunications operator, a 30% share of Ghana Telecom was sold for US$ 38 million to a consortium led by a strategic foreign investor, Telecom Malaysia Bd. (BMI-T 1998)(cf. Annex 3-1).

Other measures that significantly changed the ICT sector development of the country were the creation of a duopoly by licensing a second national network operator, as well as

- the liberalisation of value added services, such as mobile cellular telephone services, data transmission, paging and payphones;
- the creation of the possibility for large corporate users to develop their own private networks; and
- the establishment of a regulatory agency for the sector.

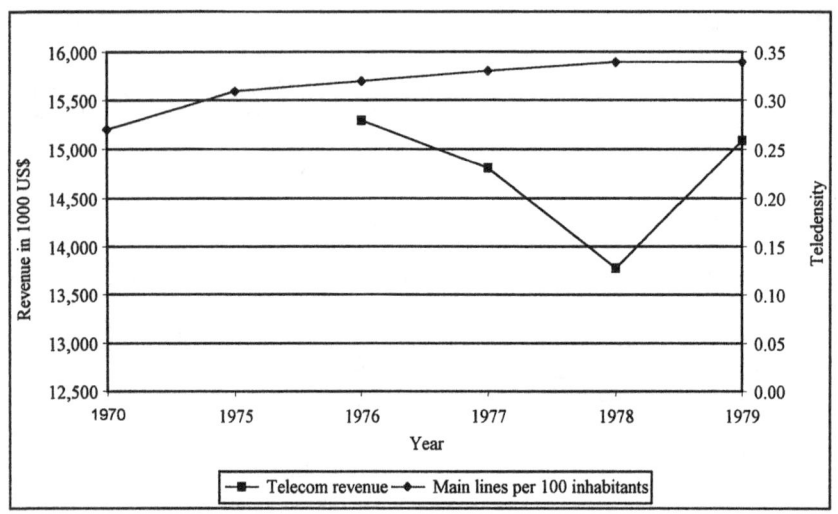

Figure 3-3: *Development of fixed main lines per 100 inhabitants and revenues in US$*
(data for 1970-1976 not available) of the Ghanaian PTO during the 70s
(Data source: ITU 2000)

In line with issues raised in Section 3.1.2, the National Communications Authority (NCA) was established in 1996 to supervise the implementation of the measures mentioned by promoting fair competition and protecting operators and consumers. Moreover, the NCA was to supervise the fulfilment of the preconditions linked to the licence agreements that were set up in order to reach the goal of improved accessibility (MINISTRY OF TRANSPORT AND COMMUNICATIONS 1999).

3.2.1.2 Market Players and Infrastructural Developments on the National and Sub-national Level

The political and institutional framework induced developments that at the first glance are evident if one considers the number of market participants. Although Ghana Telecom still controls the vast majority of the market, the additional agents will shortly be introduced (cf. Figure 3-4).

On the fixed line market, ACG Telesystems Ghana is the second national network operator. It is owned by the Ghanaian Petroleum Company (GNPC), Western Wireless (USA) and ACG Telesystems (USA) and operates under the trade name WESTEL (BMI-T 1998). Together with Ghana Telecom, WESTEL will be the only fixed line operator in Ghana until the year 2002 when the market will be opened completely. This agreement could be achieved on the basis of an attached precondition stipulating that WESTEL would install 100 pay-phones and 50.000 fixed lines between 1998 and 2001 investing US$ 30 million.

Due to the fact that the company "has yet to fully organise itself to compete seriously" (FREMPONG, ATUBRA 2001, p. 207) and is facing difficulties in settling interconnection agreements with Ghana Telecom, the number of residential and business lines implemented by WESTEL is still relatively low. As a matter of fact the WESTEL network – by mid 1999 – consisted of only 50 payphones installed in the Accra-Tema conurbation, plus a low number of lines for high potential business customers such as Ashanti Gold Field Corporation and the GNPC (BMI-T 1998, GHANAWEB 1999). Another reason that affected the competitiveness of the company is that it operates – on the end user level – on wireless technology which, in 1999, required a prohibitive initial investment in hardware and subscription fee of US$ 260 per line.

Based on a similar technology, the wholly Ghanaian-owned company CAPITAL TELECOM was established in 1994[12] to primarily serve the rural areas in the South of the country. The company provides rural communications using a multi-access WLL radio system with a potential capacity of 1.000 subscribers per hub. Seven hubs were planned to be operational by the end of 2000. So far, three of the WLL switches have been implemented in the Southern part of the country serving around 300 customers. This somewhat disappointing performance might, again, be due to the high installation cost and monthly subscription fee (cf. Annex 4-1).

By the end of 1997, the fully liberalised mobile services market covered the major cities in the Southern part of the country and served approximately 20.000 subscribers (BMI-T 1998). Due to long waiting periods for fixed lines and the enormous demand for telecommunication services, this figure increased to approximately 40.000 in May 1999 (GHANAWEB 1999; FREMPONG, ATUBRA 2001). This is despite the fact that connection and call charges, line rental, and the costs

---

12 Capital Telecom did, however, not begin operations until February 1997.

for the handset were relatively high[13]. However, ongoing changes are expected to promote the cellular sector even further.

Firstly, the influx of mobile receivers from the US and Europe as well as the introduction of a pre-paid system significantly lower entrance costs for end users. Secondly, increased competition and a decrease in costs are expected if the three operational mobile services, Mobitel, Spacefon, and Celltel face competition from the cellular phone branch of Ghana Telecom (OneTouch) from 2001 onwards. The licence awarded to Ghana Telecom will authorise the company to attract up to 100.000 subscribers in all 10 administrative regions in Ghana. It is then expected that the more or less metropolitan market will expand to the more rural centres of the country.

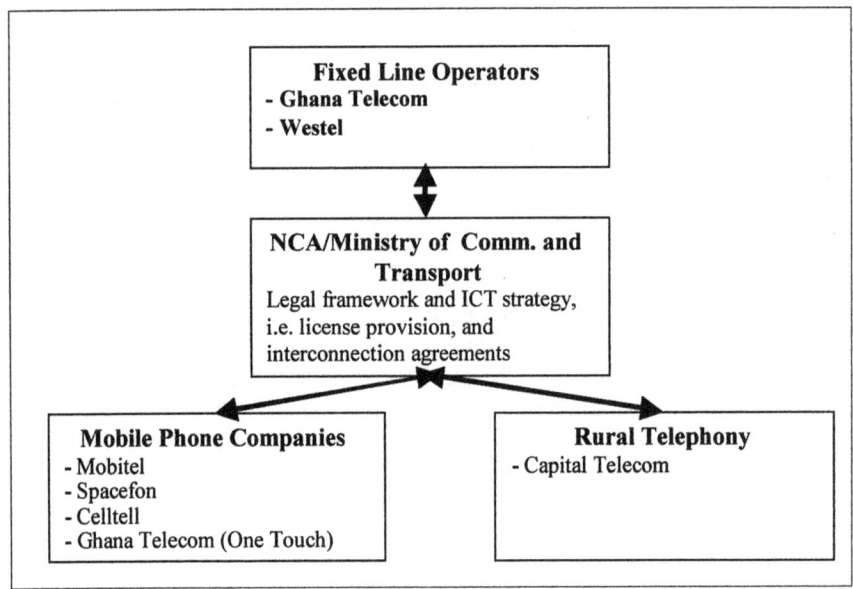

Figure 3-4:   The actors within the Ghanaian telecommunications sector (cf. FREMPONG, ATUBRA 2001)

Another – certainly somewhat different – market player is the NCA itself. Although it has the authority to make regulations on rules and guidelines on tariffs, terms and conditions for interconnectivity and technical standards, it is "grap-

---

13   Connection charges: US$ 40 to 150, monthly subscription US$ 22 to 25, call charges US$ 0.20 to 0.50.

pling with the thorny problem of putting critical structures in place to regulate and manage the sector efficiently" (FREMPONG, ATUBRA 2001, p. 205). The authors also argue that the organisation was not able to provide comprehensive regulation and by doing so weakened its own hold on the operators.

Infrastructural Developments

Despite the regulatory problems mentioned, the changing sector framework seems to have enhanced the populations' access to telecommunication services: Figure 3-5 shows how the overall number of main lines has tripled since 1990. From what was mentioned beforehand, this is due less to market liberalisation and the new market players than the network expansion by the former incumbent. The latter's motivation to expand its infrastructure is due to the aim of insuring the companies hold on the market but also due to an important regulatory agreement: Ghana Telecom retains limited liability over a 5-year period but is required to provide around 225.000 main telephone lines between 1998 and 2002, investing more than US$80 million (FREMPONG, ATUBRA 2001).

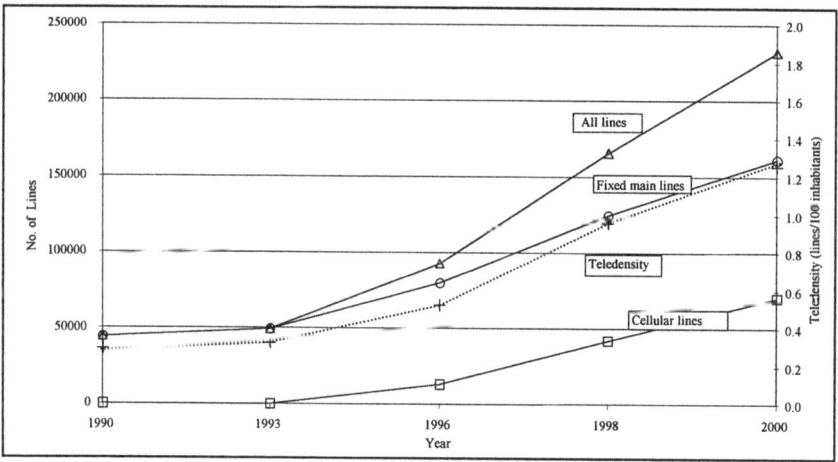

Figure 3-5: Development of fixed lines, cellular subscribers, and teledensity in Ghana between 1990 and 2000 (Data source: BMI-T 1998; FREMPONG, ATUBRA 2001; GHANAWEB 1999;ITU 1998)

Furthermore, Figure 3-5 shows that the overall increase of operational telephone sets is based on the rapid developments on the mobile phone market. Its increase in market share clearly indicates that cellular mobile services cannot be seen as a complement to the fixed line market but often compensate for long waiting periods for fixed phone lines (GHANAWEB 1999).

One negative that results from rapid fixed line and mobile phone market growth is that the switching capacity of Ghana Telecom was not expanded relative to market growth and is, hence, not sufficient for the current demand: "this appears to have affected the ability of the company to give more capacity to other competitors as well as keeping to its mandatory obligations" (FREMPONG, ATUBRA 2001, p. 205). Consequently, calling a mobile phone subscriber through the Ghana Telecom Network connection is – particularly during day times – said to be a difficult if not an impossible task. FREMPONG and ATUBRA (2001) use this example to underline the problems of the NCA and, with regard to these problems, even talk about a trade war between the mobile phone operators and the former monopolist.

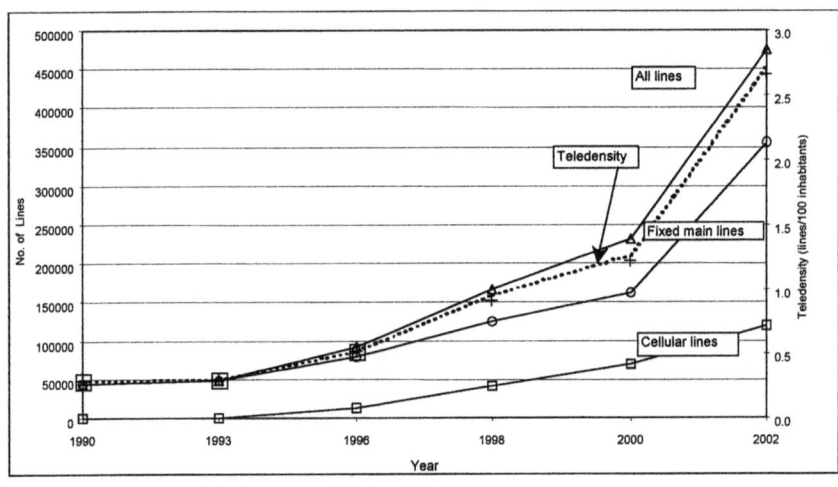

Figure 3-6: *Development of fixed lines, cellular subscribers, and teledensity in Ghana between 1990 and 2002 (Data source: ITU 1998; GHANAWEB 1999; BMI-T 1998)*

The aforementioned overload of the existing switching capacity as well as the lack of regulatory capacity makes it hard to believe that the license preconditions will be met by the year 2002. Figure 3-6 does project the situation if the license conditions were met and also reflects the cellular market to the same year, assuming a rather conservative growth rate in this segment. In light of the 100.000 sub-

74

scribers targeted by Ghana Telecom over the next couple of years, the total number of phone lines will potentially increase from the current approx. 170.000 to around 475.000 in the year 2002 (BMI-T 1998; GHANAWEB 1999).

The next section will examine whether the improvements that have been achieved so far also benefit rural areas and the overall extent to which these areas are connected to the telecommunication infrastructure.

Telecommunication Infrastructure on the Regional and the District Level

The map produced (cf. Figure3-7, also Annex 3-2 for detailed figures) shows large differences in the number of main lines between the regions. While the Greater Accra Region accounts for nearly 80% of the country's fixed telephone lines, the other nine regions range between 0.3% and 8.6% of all lines. Not surprisingly, the teledensity follows the same spatial pattern. It is highest in the Accra conurbation (4.89) and decreases rapidly to 0.39 in Ashanti and further down to less than 0.1 in the Northern, Upper East, and the Upper West Region. All in all, the major economic centres of the country, namely the Accra-Tema conurbation, Kumasi, Cape Coast, and Sekondi-Takoradi account for nearly 90% of all Ghanaian telephone lines.

To draw a more precise picture of the intra-regional distribution of telephone lines, the map (Figure 3-7) also indicates that the regional capitals account for between 48% and 100% of all fixed lines within their respective region (for details cf. Annex 3-2).

Although no detailed data are available to show the current spatial pattern of the cellular phone infrastructure, the map points out that network coverage is, again, restricted to the same urban centres but includes Obuasi, Ghana's most important gold mining town (BMI-T 1998; GHANAWEB 1999).

The analysis of the infrastructural situation at the district level shows that only 49 of the 110 district capitals were connected to the national telecommunication network in 1998 (GHANA TELECOM 1998). There are, however, districts in which other communities than the district capital have access to the national grid. Figure 3-8 shows the distribution of districts that obtain at least one community connected to the Ghana Telecom network. Figure 3-7 indicated the serious divergence in telecommunication infrastructure between the centres and the periphery on various scales and made clear that service supply follows the urban and administrative hierarchy. On the district level, however, no such pattern can be observed (Figure 3-8). Despite the lack of data and the possibility to visualise such patterns, strategic interests (towns at the border of the country) and the possibility of combining telecommunications with other forms of physical infrastructure, such as roads and electricity, might be the most important factors in determining

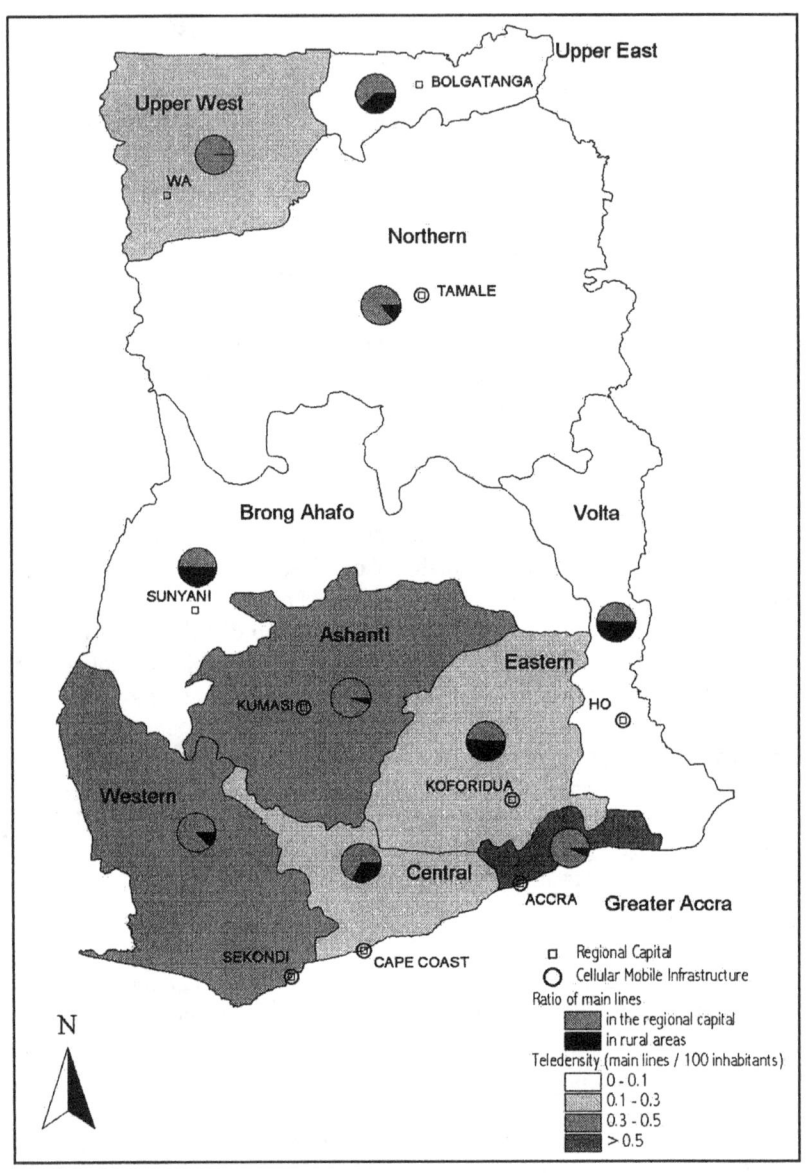

Figure 3-7: Teledensity in the administrative regions and urban-rural disparities (Data source: GHANA TELECOM 2000; for detailed figures cf. Annex 3-2)

whether services are in place or not. This might once again underline that the diffusion of services is not primarily linked to market mechanisms and indicates the existing market failure within the telecommunications sector.

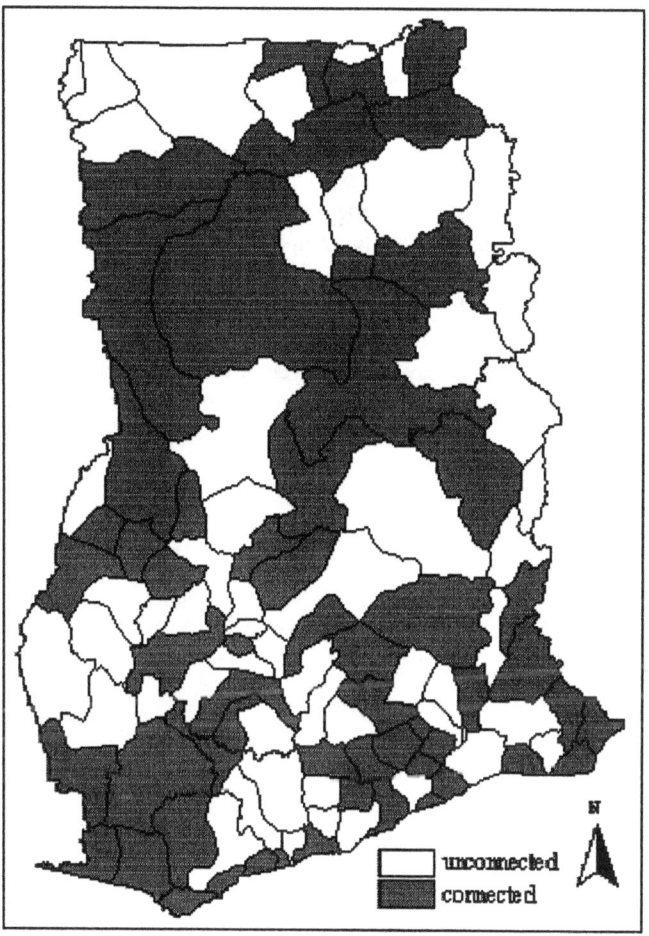

Figure 3-8: Districts in Ghana that are connected to the Ghana Telecom Network (Data source: GHANA TELECOM 2000)

3.2.2 Conclusion

Chapter 3 described the most important trends in the telecommunication sector adapted by low-income countries in general and Ghana in particular. From the assessment of the country's infrastructure we learnt that since the sector policy was introduced, rapid developments in infrastructure provision have taken place.

This resulting sectoral increase was, however, not only due to the privatisation and the liberalisation of the market, but mainly due to the target of gaining market power and expansion obligations set up by the government. As a matter of fact, the former monopolist was able to strengthen its position on the fixed line market, which in light of weak competitors, an insufficient switching capacity and an inadequate regulator, certainly hampered the overall performance of the sector and customer satisfaction (cf. FREMPONG, ATUBRA 2001).

Apart from the fact that by far not all capitals of the second administrative level of the country are integrated into the network , the weakness of the regulator does raise question marks over the universal access target formulated in the ADP. The question of how this institutional and infrastructural service supply situation is expressed at the local level and how the supply is used leads us to the next chapter.

4 Access to and Use of Public Telecommunication Services – An Example from the Southern Volta Region

In Chapter 2 the major diffusion processes and beneficiaries from ICT use were discussed, focusing on low-income countries in particular. Chapter 3 not only referred to the specific problems of Sub-Saharan Africa in terms of ICTs, but also gave an overall view of the general trends in telecommunication sector reform in the region. It also pointed out the current sector developments and their infrastructural consequences in Ghana. It became clear that improvements have not yet trickled down to the level of rural areas.

By looking at three neighbouring rural communities, one of them allocating public telephone facilities, supply and demand for the services will be assessed. In doing so, we will be able to answer the question of who is actually using the services, how this happens, and what benefits can be expected if telecommunications are available. To achieve these tasks the following steps are carried out.

First of all, the supply of telecommunication services to the rural population needs to be shown. The underlying hypothesis in this respect is that service provision is based on limited infrastructural resources and requires entrepreneurial creativity on the part of the population (Section 4.3.1).

Second, it is necessary to find out what proportion of rural households actually demand the services within each of the three communities and how this demand can be characterised in terms of its intensity (Section 4.3.2.1, 4.3.2.2).

Third, one needs to find out who uses the services. This requires the identification of the household characteristics that explain whether individual household members use telecommunications or not. If they do so, the impact of these determinants on the intensity of use will be also be determined.

Methodologically, the first two steps are based on descriptive analyses of primary data. The third, however, will be tackled in more detail. Bivariate methods will give a first idea of the interdependence between households attributes and the use of telecommunication services. Applying multivariate regression analyses allows us to show this interdependence from a more synoptical perspective. It also allows insights into the direction and magnitude of the impact of the household-related determinants on both, the use of telecommunication services and the intensity of this use (cf. 4.3.3).

In order to understand the setting in which the analyses were carried out, this chapter will, however, begin by describing the surveying process and tools with which the primary data was assessed (Section 4.1). This includes the need to briefly illustrate the process that led to the selection of Akatsi Town and two other communities in the Southern Volta Region in Ghana. The major political and socio-economic characteristics of the latter will also be portrayed (Section 4.2).

4.1 Overview of Survey Design and Assessment Tools

As already mentioned, telecommunication services in rural areas of low-income countries are usually provided in the form of public phone booths or telecommunication centres. Individual access is scarce and mainly subject to infrastructural constraints, i.e. the lack of telephone lines and/or switching capacity (cf. BERTOLINI ET AL. 2000).

Bearing this in mind, the survey location was selected according to the following criteria:
- Public access to telecommunication services should be extant.
- The location should be rural in terms of the predominant economic activities, which include agriculture, rural industry and services, as well as trade in agricultural products (cf. THE REPUBLIC OF GHANA 1995).

Apart from these factors, practical criteria such as the availability of secondary data and current household lists were considered.

In view of the first two criteria, the analysis of secondary data produced in Chapter 3 limited the choice to those Ghanaian small and medium-sized towns which are not located within the catchment area of the major urban centres but can be regarded as the service centres for their rural hinterland. Together with local researchers and consultants, the issues of practicability mentioned above were discussed and eventually Akatsi Town within the Akatsi District in the Southern Volta Region was selected as the survey site (cf. Figure 4-3).

Two further communities were included in addition to Akatsi Town itself. These two communities which are located around Akatsi Town and had no telecommunication services in place, were chosen in order to examine the reach of the Akatsi Town services. This selection was led by the concept of universal access: this concept acknowledges the fact that the attainment of telecommunication services on a personal or household level "is not yet economically nor technically feasible for many developing nations" (ITU 1998b). In fact, access is provided by publicly accessible telephone facilities. However, policy makers in different countries target different options to provide universal access (cf. Table 4-1; also Chapter 3).

To most appropriately meet our purposes, the selection of the communities was guided by the idea that one of the survey communities should be within and one outside the universal access area. The latter was defined as the area from which the telecommunication services can be reached in less than 30 minutes walking distance (cf. Table 4-1). In concrete terms, apart from Akatsi itself where the services are in place, two other communities (Agbedrafor, 1 km away from Akatsi and Gefia, 5 km away from Akatsi) were covered by the survey (cf. Figure 4-1).

Criteria	Minimum Requirement	Example
Population	One publicly accessible telephone for every permanent settlement of a population $\geq$ x	In Ghana, defined as a minimum of one telecommunication facility in every community of more than 250 inhabitants.
Distance	One publicly accessible telephone within x kilometres	In Burkina Faso, defined as a minimum of one telecommunication facility within every 20 km.
Time	One publicly accessible telephone within x minutes	In South Africa, defined as a minimum of one telecommunication facility within a 30 minutes travelling distance.

Table 4-1: Different definitions of universal access (ITU 1998b)

A variety of tools were developed and applied to generate a picture of the survey location that not only covers telecommunication issues but also assesses its socio-economic situation. They consisted of expert or key informants as well as of household interviews (cf. THE PROBE TEAM 1999; SCHNELL, HILL, ESSER 1999).

Experts or key informants, according to MEUSER and NAGEL (1991), are such persons that have privileged access to information about groups of persons and the decision processes of these groups. The research utilises this exclusive experience and the knowledge such persons have due to their distinct position within the respective community (MEUSER, NAGEL 1991; THIEM 1998). As well as members of the local administration and traditional authorities, the owners and managers of telecommunication centres were interviewed. With the *community level expert group interviews* that were carried out on the basis of a semi-structured questionnaire, information could be assessed that related to the socio-economic situation of the communities, their infrastructural endowment, as well as the opinion of the experts about the (potential) role of telecommunication services for their communities.

The *telecommunication centre expert interviews* together with secondary data available from the *Technology Assessment Project (TAP), University of Ghana,* mainly forms the basis for the analysis of service supply and also allowed conclusions concerning the catchment area of the services.

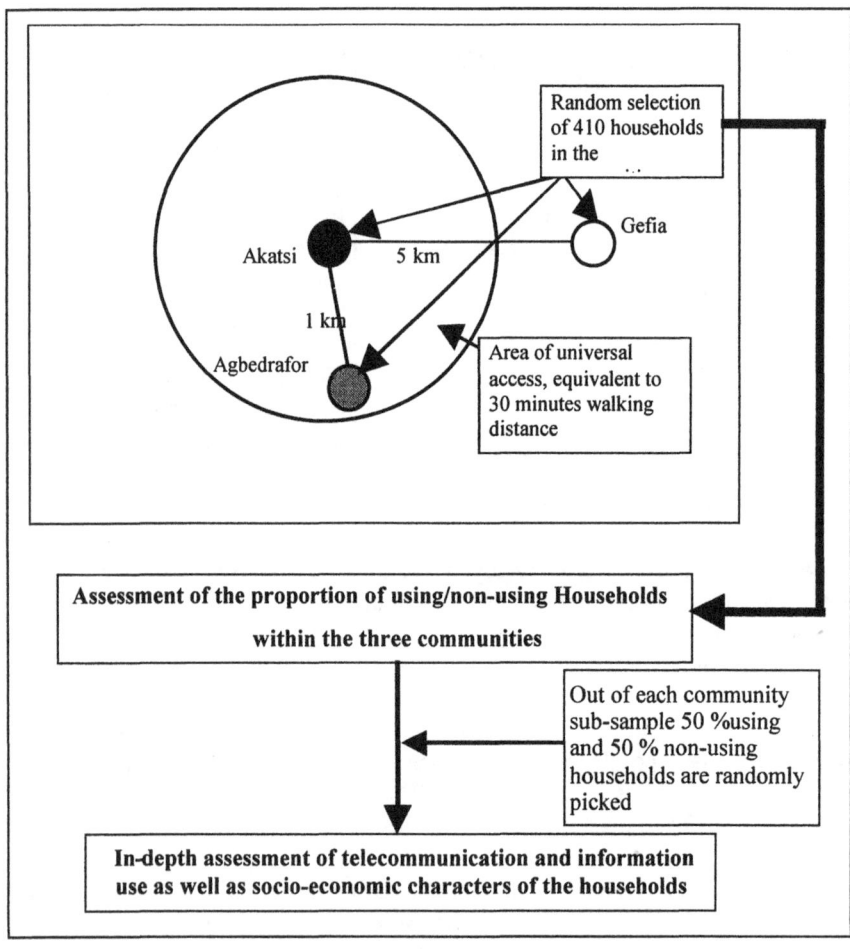

Figure 4-1: Schematic overview of the surveying process and tools

In a first round of household interviews that were based on a *standardised questionnaire*, we merely assessed whether households used the services or not. In practice 218 questionnaires could be completed in Akatsi itself and an additional 96 in each of the other communities.

This allowed a grouping into using and non-using households. Afterwards, a more detailed *household questionnaire* was applied to randomly picked members of both groups. It assessed the households' main socio-economic features and the characteristics and purposes of telecommunication service utilisation. Eventually, 170 detailed household questionnaires could be used for the forthcoming data analysis. A graphic overview of the survey design and the sampling process is provided in Figure 4-1.

An up-to-date household list, available for all three communities, built the sampling frame of the survey. Households – in line with the definition of the Ghanaian Statistical Service – were understood as a group of people "eating out of the same pot". This group is, thus, considered as an economic unit living in the same house or compound and jointly contributing to its income (cf. DICKSON, BENNEH 1995; GHANA STATISTICAL SERVICE 2000).

Due to the population's interest and its openness towards the research issue, supplementary information could be generated beyond the application of the questionnaires. Where relevant, these pieces of information were recorded and analysed in a qualitative manner.

4.2 Characteristics of the Survey Area

An understanding of household behaviour in terms of the role of telecommunication services can be provided by explaining the socio-economic and cultural situation of the population. This situation should now be related to the survey site, focusing on the specifics of the relevant administrative levels, i.e. the regional, the district, and the community level.

As far as the regional level is concerned, secondary literature and data will be utilised for a brief overview. For the district level, the Medium Term Development Plan 1996-2000 (AKATSI DISTRICT ASSEMBLY 1996) together with first-hand information from the district assembly representatives forms the information basis. The information related to the community level was generated entirely by applying the expert interviews mentioned.

4.2.1 The Volta Region

Covering approx. 20.600 km², the Volta Region ranks fourth in area among all Ghanaian regions. Located East of Lake Volta, this part of Ghana was – until the end of World War One – part of the German colony Trans-Volta, was then administered by the British and eventually joined the Republic of Ghana on the basis of a plebiscite in 1956 (DICKSON, BENNEH 1995; SCHMIDT-KALLERT 1994).

With its vast North-South extension, the region allocates the highest and the lowest points of the Republic. Most of the Northern and the Southern parts, however, are characterised by savannah areas that provide good conditions for cattle breeding and the cultivation of vegetables and tobacco. The forest zone in between the savannah plains is dominated by the fold mountains of the Akwapim-Togo Range (cf. Figure 4-2). Extensive cocoa, coffee and maize growing is the predominant form of land use in this mountainous area. In the South of the region, the delta of the Volta River and especially the Keta Lagoon are fishing grounds of national importance (DICKSON, BENNEH 1995; SCHMIDT-KALLERT 1994).

In anthropological terms, the Ewe represent the majority of the region's population. The fact that this is also Togo's major ethnic group and that the Ewe language is not related to any of the Akan languages of the remaining parts of the Ghanaian South, once again, underlines the particularities of the colonial past of this region.

In terms of population, the region experienced an increase of 1.8% between 1984 and 2000 from approx. 1.211.900 to 1.612.300 resulting in a current population density of 78 inhabitants: a figure that is almost precisely in line with the overall average for the whole country (GHANA STATISTICAL SERVICE 2000).

Box 2: The Keta Lagoon

The former harbour of Keta, an important commercial and administrative centre is located on a narrow sand bar between the Keta lagoon and the sea in the South of the Volta Region. For many years the sea has steadily been eroding the sand bar and advancing into the town. For that reason, portions of Keta are now under water and the sea is still advancing inland forcing people to leave (DICKSON, BENNEH 1995).

The main primary economic activities in the region are farming, livestock rearing, fishing and salt making. The predominant farming system is bush fallowing that mainly aims at the cultivation of food crops for home consumption and the local markets. There are few purely sale-oriented permanent cultivation practised by state and peasant farms. Apart from these occupations, small-scale industries, handicrafts, and trade are important sources of income for the region.

Alike most other parts of the country, the absence of an adequate transportation infrastructure is a great obstacle in the region. The North of the Volta region in particular has a very limited network of roads with Lake Volta forming a barrier to the Western regions. The main transport corridors are the Accra-Aflao road passing Akatsi and leading to Lomé and Lagos, the North-South road between Akosombo and Hohoe and the Ho-Denu Road (cf. Figure 4-2).

Figure 4-2: Map of the Volta Region (Source: Central Intelligence Agency 2000)

As far as the telecommunication infrastructure is concerned, the Volta Region is only scarcely connected. This especially holds true if one compares the region with other regions in the South of the country. As obvious from Figure 3-7, more than 1.000 inhabitants share one main telephone line in average. More precisely, a total of 1455 telephone lines are available for the 1.6 million inhabitants;

48.8% of all main lines are concentrated within the regional capital (cf. Annex 3-2). Figure 3-8 also indicates the concentration of those districts being connected to the network in the South of the region.

4.2.2 Akatsi District

Akatsi District is located in the south-eastern part of the Volta Region comprising an area of approx. 900 km^2 and, among others, bordering the district of the regional capital Ho in the North and the Republic of Togo in the North East (cf. Figure 4-3).

In the year 2000 around 88.000 people lived in the district (GHANA STATISTICAL SERVICE 2000). On the basis of this census data, the population increased by approximately 3.2% as compared to 1995. This increase in the district's population is, according to the AKATSI DISTRICT ASSEMBLY (1999), partly due to refugees from the Republic of Togo as well as from people leaving the environmentally threatened town of Keta (cf. Box 2).

According to the GHANA STATISTICAL SERVICE (2000), the proportion of the volatile population (the 15-64 year olds) of the district makes up around 50%. The average family size is 5-6 individuals (AKATSI DISTRICT ASSEMBLY 1996, 1999).

The district's settlement pattern is rather disperse and consists of 829 settlements of which only eight consist of more than 1.000 inhabitants. The remaining majority are farm settlements. Besides the district capital, Akatsi Town, the 2 communities that were selected, namely Agbedrafor and Gefia, belong to the group of larger settlements in the district (AKATSI DISTRICT ASSEMBLY 1996).

Population of the Settlement	No. of Settlements	Percentage
>5000	1	0.12
2501-5000	1	0.12
1001-2500	6	0.72
200-1000	51	6.16
<200	770	92.88
Total	**829**	**100**

Table 4-2: Population distribution in the Akatsi District related to the size of the settlements (Source: AKATSI DISTRICT ASSEMBLY 1996)

As for the whole region, bush fallowing farming and animal husbandry are the predominant economic activities within the district: around 75% of all households live from this occupation. On an average farm size of 2-2.5 ha they grow maize, cassava, cowpeas, pepper, tomatoes and groundnuts mainly for home consumption. The sale of surplus food on the local markets serves as supplementary income (AKATSI DISTRICT ASSEMBLY 1996, 1999). Commercial farming is scarce and restricted to tobacco and vegetable cultivation. Livestock keeping of cattle, sheep, goats, poultry, and pigs mostly exists on a small-scale level.

In its development plan the AKATSI DISTRICT ASSEMBLY (1996) defines the major problems related to agriculture and agro-processing: besides the increased rainfall variability, the lack of government support for small-scale farmers, high input costs, the lack of possibilities to market excess produce, and the limited agricultural extension system were mentioned. Furthermore, the reluctance of most farmers to the adapt modern techniques and innovative ideas was pointed out.

Commercial activities and trade are "reasonably well developed in the district, by rural Ghana standard" (AKATSI DISTRICT ASSEMBLY 1996, p. 21). This especially applies to the district capital. Akatsi Town hosts a periodic market for farming and some industrial goods that is of major importance for the district and beyond. The centrality of Akatsi Town is further strengthened by the presence of two banks.

However, a formal manufacturing sector is not developed. Ongoing small-scale manufacturing activities are mainly informal in their nature and include carpentry, blacksmithing, basketry and mat weaving. Welding, masonry, tailoring, are mostly operated as family or individual businesses (AKATSI DISTRICT ASSEMBLY 1996, 1999).

As far as the public infrastructure is concerned, the education and health sectors are of major importance. The former includes facilities ranging from numerous nurseries, primary and secondary schools, and a training centre. It is important to note that, especially in the remote areas, educational institutions suffer from a lack of physical infrastructure such as buildings and furniture as well as from a lack of qualified personnel. These problems equally apply to the health sector whose low standards were also pointed out. According to the AKATSI DISTRICT ASSEMBLY (1996) the major problem is that most people live too remote from the existing health facilities. The main cause of health problems is considered to be the poor water supply and the highly inadequate sanitation facilities on both domestic and public levels.

The decentralised administrative structure provides the major framework for the district and local authorities. The former takes the form of an assembly that represents the highest political, legislative, and executive authority in the district. It comprises the district administration and representatives from the local authorities. The Akatsi District Assembly has 54 members that include 36 elected members as well as 18 government appointed representatives. The executive duties of

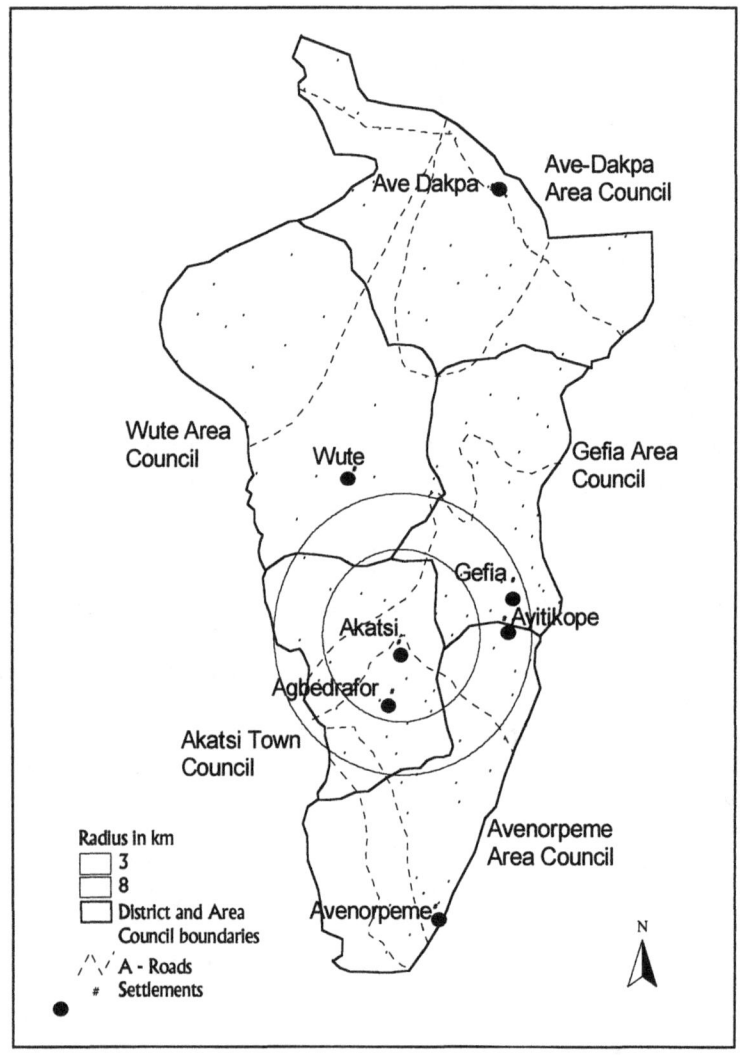

Figure 4-3: Map of Akatsi District (Source: Akatsi District Assembly 1999)

the assembly are performed by the Executive Committee. This committee does consist of a maximum of one third of the members of the assembly, it adopts measures to develop activities and executes the plans of the local authorities within the district boundaries. Moreover, it monitors the work of the district administration, which is headed by the District Co-ordinating Director (DCD). As the head of the district's civil service, the DCD oversees the running of the government organisations in the district.

The implementation of the latter takes place through towns and area councils (cf. Figure 4-3). Within the Ghanaian administration these councils represent the lowest level of local government structures. They are not independent from the District Assembly but perform important functions such as (AKATSI DISTRICT ASSEMBLY 1996):
- Establishing the building-up of the community development committee.
- Enumeration and record-keeping of rateable persons and properties in the district.
- Assistance in preventing or eradicating public health hazards.
- Discussion and resolution of local problems.

Parallel to these local government systems are institutions that evolved in pre-colonial times. The most important representatives of those institutions are the Chiefs. So, apart from the assembly member who represents the town or area council in the district assembly, the Chief is the second main executive representative of a specific town. Both functions share similar obligations in their day-to-day activities, which consist of conflict resolution, development-related issues and emergency operations. To achieve consistency between the different authorities, close co-operation and co-ordination is an important responsi-bility. Furthermore, the traditional authorities' representatives meet at district level to discuss issues of mutual importance and to co-ordinate activities in their specific re-sponsi-bilities (AKATSI DISTRICT ASSEMBLY 1996).

Overview of the District's Telecommunication Situation

Within Akatsi District the only publicly accessible telecommunication facilities are located within the capital Akatsi. There is no mobile cellular telephone service available although the Accra-Aflao Road is due to be covered by the service providers on a medium-term basis. In terms of conventional telephony, an approximate 20 lines are provided through an extension from an exchange located in the regional capital Ho. All those lines are located within Akatsi Town itself; seven of them are publicly accessible. Despite long waiting lists, this situation could not significantly be changed even though CAPITAL TELECOM (cf. Chapter 3.2.1) officially launched its WLL service in 1998. According to members of the AKATSI DISTRICT ASSEMBLY (1999), the low acceptance of the service is mainly due to the high initial and monthly fixed costs of the system: compared with the former monopolist, the new entrant asks about 3 times the connection charge and, including equipment rental, 12 times the monthly fixed costs (cf. Annex 4-1). Moreover the WLL equipment requires stable and permanent access to electricity.

4.2.3 Overview of the Communities Surveyed

The previous sections provided an overview of the regional setting and explained some of the major characteristics of the district. We will now shift to the community level and, based on information obtained from key informant interviews, try to sketch the *site and situation* of Akatsi as well as of Agbedrafor and Gefia. In this respect *site* is understood as the particular location of the specific community. Of much greater importance within this context, *situation* refers to the position of the site in relation to other settlements, roads etc. (DICKSON, BENNEH 1995).

Akatsi Town

There is no doubt that Akatsi is the central place in the district and beyond. This importance is not only reflected by its population level of just over 5.000 inhabitants, but also by the small town's economic and administrative functions and the existing physical infrastructure. This physical infrastructure mainly consists of the connection of the town to the national electricity and telephone network as well as the crossing Accra-Aflao road, which links the capitals of Ivory Coast, Ghana, Togo, Benin, and moves further west up to Lagos.

The government's decision to make the town the capital of a newly established district was a major boost to Akatsi's development. This took place as part of national local government system reform which was carried out in 1988. Apart from the fact that the town enjoyed more attention from the central government, this move created direct employment, which had an impact on the local commodity and food markets, small-scale industries and services. The latter mainly include transportation, financial, secretarial, and photocopying services, as well as legal assistance (AKATSI DISTRICT ASSEMBLY 1996, 1999).

It is, therefore, not surprising that – compared with the district as a whole – Akatsi's population only consists of 25% farming households with the remaining three quarters being mainly involved in trade, agro-processing and the civil service. Most of these households do, however, grow agricultural produce for their own consumption (AKATSI DISTRICT ASSEMBLY 1996, 1999).

The transportation services available include the national express-coach service, a couple of taxis as well as numerous minibuses. On a continuous basis, they connect Akatsi to the regional capital Ho in around one and a half hours and to the industrial town of Tema and Accra in two and a half and three hours respectively.

Although nurseries, primary, and secondary schools are distributed all over the district, the facilities located in and around Akatsi are better equipped. The only professional training centre in the district is also allocated in the district capital. Similarly, the health sector situation is above the district's average and includes one public and two private clinics, and a maternity home (AKATSI DISTRICT ASSEMBLY 1996, 1999). The supply of safe water in Akatsi Town is above the district's standard due to safe boreholes and the widespread application of water harvesting methods.

The institutional structure and its functions have already been explained above. Nevertheless, there are arrangements that are not necessarily government-related, but play an important role in the every day life of the population. The most important amongst these institutions are trade and marketing associations, co-operatives, and farmer extension groups. Furthermore, there are savings and loan groups such as the Rotating Savings and Credit Associations (ROSCAS) and the Enhancing Opportunities for Women in Development Group (ENOWID). Additionally there are traditional systems of mutual aid such as the loan system SUSU and reciprocal labour arrangements.

Agbedrafor

Agbedrafor is located around 1 mile away from the centre of Akatsi Town. The proximity to the district capital also enables the population of the community to take advantage of Akatsi's centrality functions without being forced to travel huge distances: the distance to Akatsi Town is not more than one and a half kilometres and requires a 30 minute walk (AGBEDRAFOR 1999).

Despite this proximity and the relative importance of the community with slightly less than 1.000 inhabitants, the infrastructural situation is – according to the interview partners – far from satisfactory. Although the Ghanaian Electricity Corporation (ECG) asked the villagers to set up poles for the installation of electricity lines, a collective decision was taken to instead invest the money required in a second borehole to secure access to safe drinking water. As indicated above, there is no telephone connection to the community and there is no information as to whether the telecom providers are planning to implement such a connection.

The road that connects the community to the Akatsi main road (Accra Aflao Highway) is not tarred and is in a bad state (AGBEDRAFOR 1999).

Most households in the community live from farming staple products and vegetables with some also being involved in civil service, petty trading and craft-making such as basket weaving. Basically all economic activities are directed towards the nearby Akatsi Town. As there is no regular transport to the district capital, people mostly walk to Akatsi, carrying the goods they sell. As raised in the interview, some farmers in the community own small cocoa farms in the north of the Volta Region where they frequently travel in the harvesting season. Cocoa harvesting in the North is also an income source for pupils and students during their school holidays.

The community has one primary school. Secondary schools and health facilities do not exist in the community as such but are easily accessible in Akatsi. Akatsi, once again, is the centre for most people in terms of their associations, co-operatives and loan groups. Apart from the local authority, only a women's loan group (ENOWID), SUSU collection, and mutual help in farming labour are more or less formally arranged within Agbedrafor itself (AGBEDRAFOR 1999).

Gefia

Gefia is located further north of Akatsi on the vertex of a hill. Gefia is the capital of one of four area councils in the district (cf. 4.2.2) and, with around 1.000 citizens[14], also one of the most important communities in the district. This importance is, however, not reflected by the small town's infrastructural and economic situation: There is neither electricity nor a telephone connection in place. The interviewees (GEFIA 1999) strongly emphasised the problematic water supply. The borehole stopped working and – apart from limited water harvesting – a dam is the most important source of drinking water. The administrative centrality of Gefia is not reflected by the town's economic functions. The market, post office, and banks are located in Akatsi Town which is approximately five kilometres away and can be reached in an one and a half hours walk on gravel road, by using the irregular public transport system or by hitch hiking.

As in Agbedrafor, households mostly live from farming. Few other employment opportunities are provided by the local government. Additionally, some households are involved in small-scale trading and the production of Akpeteshi, the local palm-oil-based brandy.

---

14   It was not possible to obtain consistent answers on the actual number of Gefia's population.

In line with its administrative function, primary and secondary schools are in place, although badly equipped. A nurse and a midwife are currently caring for health-related issues in the local health facility which is allocated in a small clinic in the centre of the community.

Outside the public administration, a women's loan group (ENOWID), SUSU collection, and mutual help in farming are the only formal institutions in Gefia (GEFIA 1999).

4.3 Access to and Use of Telecommunication Services in Akatsi

4.3.1 Telecommunication Services – the Supply Side

As already mentioned, the forthcoming supply analysis is led by the hypothesis that the provision of telecommunication services is based on limited infrastructural resources. To verify this hypothesis, the prevalent information and communication infrastructure will be assessed for the target area. Furthermore, the means by which the communities have access to telecommunication services will be characterised. As micro-enterprises are the most important of these interfaces, not only cost information (as for the existing public phone box) but also the managers' perception of the customers and business-related information will be analysed.

4.3.1.1 General Overview

According to the community and district leaders, the availability of cheap and reliable access to telecommunication services is not satisfactory in each of the three communities. Whereas Gefia and Agbedrafor do not have a local connection to the telecommunications network at all, service provision in Akatsi Town is said to be expensive and unreliable. Again, according to the experts this is despite the high, but as yet unmet, demand that is expressed by the 250 households and small firms in Akatsi Town alone that have applied for a Ghana Telecom connection. Due to the fact that there is neither a switch nor a repeater station in place, the number of lines extended from the regional capital Ho stagnates at around 20 (AKATSI DISTRICT ASSEMBLY 1999).

However, within the last five years changes have occurred as far as the public accessibility of these scarce resources is concerned. Seven of the existing Ghana Telecom connections have been converted to one public card telephone and six privately run communication centres. Apart from those lines, only public institutions and a few larger enterprises are connected to the Ghana Telecom network.

Additionally, there is a CAPITAL TELECOM phone publicly available in the *Black Cat Communication Centre*, which is part of a hotel. This centre and the *African Arts and Communication Centre* could not be part of the *TAP* survey and neither could the *communication centre expert questionnaire* be applied (cf. Section 4.1). In the case of the former, this is due to the fact that the facility is only very rarely used by the local population but instead serves the small hotel and its guests. In the case of the latter, information could not be obtained because the facility was not open for a substantial length of time and its future operation could not be clarified. The public card phone available at the post office will also not explicitly be dealt with. It will play a role as its presence relates to the operation of the telecommunication centres. We are therefore left with six publicly accessible telecommunication facilities, which were all implemented after year 1995 (cf. Table 4-3) and will now be introduced in more detail, focusing on the following in particular:

- General information about the enterprise establishment, its financing, and management.
- The staffing and ownership situation.
- The range of services supplied to the public.
- The costs of the telecommunication services offered.
- Specific success stories and problems that have occurred since the establishment of the enterprise.
- The characteristics of the clients served by the facilities; this also includes an attempt to define the catchment area of the services offered.

Type of Access and Name	Service Provider	Year of Establishment
Akatsi Premier Communications Centre	GT	1996
Zutako Communications Centre	GT	1997
Zutako Communications & Business	GT	n.a.
West Falia Communications Centre	GT	1998
Benak Trading and Communications Centre	GT	1997
Black Cat Communications Centre	*CAPITAL TELECOM*	*1998*
African Arts and Comm. Centre	*GT*	*n.a.*
Public Phone Box	*GT*	*1995*

Table 4-3: Public access facilities in Akatsi Town. The facilities that could not be included in the survey are shown in italics (TAP 1999)

4.3.1.2   Public Call Offices in the Survey Area

The five public call offices (PCOs) that could eventually be included in the survey (cf. Table 4-3) do have rather diverse characteristics. They all have in common, however, that they are operating on the basic telecommunication infrastructure that was provided by Ghana Telecom from 1996 onwards. As already stated, no technologically sophisticated infrastructural means were used: analogue copper lines that extend from a switch in the regional capital Ho 70 km away connect the communication centres to the national network.

Box 3: Sénégal's Télécentre-Privé

A 'télécentre-privé' is a privately owned small enterprise that provides access to telecommunications on the basis of a licensing-contract with SONATEL. To be able to open such a centre, the following preconditions have to be fulfilled:
* *The centre must be open to the public.*
* *The entrepreneur must have a trading licence.*
* *A shop of 12m² minimum must be available to install the centre.*
* *The following payments must be secured:*
   - a deposit per line of around FCFA 500.000 for a centre in Dakar and FCFA 300.000 for any other areas,
   - a connection charge per line of about 67.500 FCFA
   - a share of 60 FCFA per unit sold.
This concept led to higher connectivity especially in the suburbs of Dakar and smaller towns. As competition increased, many entrepreneurs, especially in the urban areas, reduced their charges from 100 FCFA to 80 FCFA and hence reduced their profits from 40 to 20 FCFA per unit. (cf. BERTOLINI 2000).

In comparison with other countries, Ghana has so far not embarked on a common scheme for the establishment of telecommunication centres. As a result, the centres are run like private enterprises that resell the service and are not enjoying lower tariff rates or other benefits of a reseller. The latter could, for instance, be very successfully established in the Senegalese *Télécentre Privé* Scheme by SONATEL (cf. Box 3). It was difficult to obtain concrete answers from the owners and managers concerning how much has been invested in monetary terms to establish the enterprise. The same applies to the question about the source of finance that was tapped in order to set up the enterprise. The owner and manager of *Akatsi Premier Communication Centre* and *Benak Communication Centre* respectively indicated that the major financial input was generated by own funds. The former was also supported by a bank loan and relatives (cf. Annex 4-2/Table 1).

The question of why the enterprises were established at their particular location can, generally speaking, be answered by the observation that all centres are situated around the business centre, i.e. the market of the town. More specifically, *Akatsi Premier* was established next to the owner's house on a side road, whereas *West Falia* and *Benak* were intentionally set up along side the Accra-Aflao Road (cf. Annex 4-2/Table 2).

Apart from Akatsi Premier Communication Centre, which is managed by its female owner, all businesses are run by managers employed by the male owners of the premises. Apart from the managers, three of the centres have another employee; the *West Falia* and *Benak* centres are officially run by just one person (cf. Annex 4-2/Table 3).

The public card phone seems to be competitive in relation to the allocation of the existent PCOs: it is situated at the post office in the centre of town along side the main road. However, the range of telecommunication services supplied to the public is in favour of the PCOs (AKATSI DISTRICT ASSEMBLY 1999). Apart from outgoing calls they are able to handle facsimile services, incoming calls and the manager or employee can record messages for the PCOs' clients. As far as the telecommunication services are concerned, all centres have just one telephone connection – apart from the *Akatsi Premier* which has two. Two centres use their telephone connection to provide facsimile services.

The opening hours do not vary significantly between the centres and are basically in line with the town's general business hours (cf. Annex 4-2/Table 4 and 5).

Name of enterprise	Rank of importance of			
	outgoing calls	incoming calls	photo-copying	other services
Akatsi Premier Communications Centre	1	2	3	0
Zutako Communications Centre	2	3	0	1
Zutako Communications & Business	1	3	0	2
West Falia Communications Centre	2	1	0	0
Benak Trading and Communications Centre	1	2	0	0

Table 4-4: Ranking of importance of the services offered (Data source: TAP 1999)

Three PCOs supply services beyond basic telecommunications. In the case of *Akatsi Premier* this is photo-copying service, whereas both *Zutako* Centres engage in petty trading. In terms of economic importance, only the manager of the *Zutako Communications Centre* claims that trading activities account for more revenue than the supply of telecommunication services (cf. Table 4-4).

As far as the services offered are concerned, Table 4-4 reflects the importance of the different telecommunication services amongst each other and how these relate to the other economic activities of enterprise.

The owner and managers of the centres were also asked questions concerning the success of their operation as well as their business strategies.

The success of the enterprise is reflected by the average number of daily clients and the revenue generated. Table 4-5 shows that, on average, 15 to 85 clients per day use the facilities generating a daily revenue between ¢17.000 and ¢50.000. Table 4-5 also shows that the vast majority of the calls are trunk calls, i.e. national calls outside the Ho area.

Name of enterprise		Int. direct dialling (IDD)		Local calls		Trunk calls		Approx. average revenue per day
	All	In	Out	In	Out	In	Out	
Akatsi Premier Communications Centre	17	n.a.	n.a.	n.a.	n.a.	15	2	¢20.000
Zutako Communications Centre	15	2	1	0	0	10	2	¢17.000
Zutako Communications & Business	42	n.a.	n.a.	2	2	35	3	¢45,000
West Falia Communications Centre	85	3	5	5	2	40	30	¢50.000
Benak Trading and Communications Centre	26	n.a.	n.a.	3	3	15	5	¢25.000

Table 4-5: Number of customers and approximate average daily revenue from telecommunication services (Data source: TAP 1999)

In terms of their strategy to attract customers, all managers pointed out the importance of providing their customers with a courteous reception. Only the manager of the Akatsi Premier PCO emphasised the importance of providing comfortable seats and recent newspapers to their waiting customers. The fact that all interviewees pointed out that they would like to expand their services may indicate that they would like to acquire more telephone lines. Despite the constraints concerning this expansion in terms of the limited line capacity available, the respondents are all satisfied with the services of Ghana Telecom (cf. Annex 4-2/Tablec 6).

Finally, the interviewees were asked to answer open questions about success stories they are able to tell and the major problems they are facing related to the enterprise. Hardly any answers were given concerning success stories. It was merely pointed out by the manager of one communication centre that the centre is helping people to organise their business activities more easily (cf. Annex 4-2/Table 7).

A greater variety of answers could be generated by the last question. One owner complained about the lack of co-operation between the communication centre owners and the fact that he has to travel as far as Aflao in order to settle the monthly telecommunications bill. Most of the complaints centred around the customers. According to the managers they often begin serious discussions about their bill. From the businesses' point of view, the fact that customers make calls that they are eventually not able to afford as well as their unwillingness to pay for incoming calls – which creates high opportunity costs for the enterprise – are problematic (cf. Annex 4-2/Table 7).

Although these problems do not occur with respect to the public card phone, the community experts as well as the communication centre owners point out that, generally speaking, most people prefer the PCOs. This is mainly due to the fact that (AKATSI DISTRICT ASSEMBLY 1999):
- The initial investment in a telephone card is, with a minimum of ¢6.000, rather high and calling cards are at times difficult to acquire.
- The public phone booth cannot receive incoming calls and messages cannot be left.
- The public card phone is often out of service and is very unreliable even during calls.

Box 4: E-mail and Internet service constraints
The largest copy shop in town utilized a PC of the latest generation to design all kinds of posters, letters, adverts and announcements. Asked why he did not use and provide Internet or email services, the owner pointed out that he would immediately do so if he had a telephone connected to his shop.

Although some private secretarial service providers and one copy shop use personal computers in Akatsi, none of them are connected to the telecommunication network (cf. Box 4). The machines were mainly utilised to design and print official letters, funeral announcements or other events important to people's lives. One secretarial service also provided word processing and type writing training on its DOS based PC.

4.3.1.3 Cost of the Services

Surprisingly for such a limited number of competitors, the prices the centres charge for their services vary rather significantly although all but one PCO produce a detailed list of service charges to their clients. As shown by Table 4-6 the outgoing calls and fax services for the first 3 minutes of a trunk call range from 1.200 to 1.350 and from 600 to 1.100 for local calls respectively. The latter, however, include calls within the Ho exchange, which means that local calls are not necessarily directed towards respondents within Akatsi Town (AKATSI DISTRICT ASSEMBLY 1999).

Name of enterprise	IDD	Trunk call		Local call	
	Cost/min.	first 3 min.	extra min.	first 3 min.	extra min.
Akatsi Premier Communications Centre	n.a.	¢1200.00	300.00	¢600.00	¢100.00
Zutako Communications Centre	n.a.	¢1200.00	400.00	¢700.00	¢300.00
Zutako Communications & Business	n.a.	¢1320.00	400.00	¢1100.00	n.a.
West Falia Communications Centre	¢4000.00	¢1200.00	400.00	¢600.00	¢200.00
Benak Trading and Communications Centre	n.a.	¢1350.00	450.00	¢600.00	n.a.
Ghana Telecom	¢2000.00-3000.00	¢200.00[15]	200.00	¢200.00[16]	n.a.

Table 4-6: Price structure of telecommunication services in Akatsi (Data source: GHANA TELECOM 2000; TAP 1999)

---

15 Price depending on distance and time of day but ranging from ¢200 per minute (national call over 80 km between 6a.m. and 6p.m.) up to ¢200 for 5 minutes (national call within 32 km between 6p.m. and 6a.m.).

16 Price depending time of day, ranging from ¢200 per 4 minutes (between 6a.m. and 6p.m.) up to ¢200 for 5 minutes (between 6p.m. and 6a.m.).

The cost of a call is eventually calculated by multiplying the time on the phone, measured by a stopwatch, with the charge specified for the destination of the individual call.

Table 4-6 also indicates that the clients are forced to purchase telecommunication services at higher per unit prices than Ghana Telecom charges people with residential access. Either they have to buy phone cards or they pay the tariffs of the telecommunication centres. The latter face the same conditions as residential subscribers (see Table 4-6) and, hence, generate their profit by demanding higher rates from their customers. For incoming calls the customers are usually charged a rate of around ¢600 per call which can be augmented as the length of the individual call increases (AKATSI DISTRICT ASSEMBLY 1999).

4.3.1.4 The Customers

Another task of this section is to outline the perception of the communication centre managers of their clients and to try to define the catchment area of the services offered.

Name of enterprise	Customer type			Approximate proportion (%) of	
				business-related calls	regular[17] business clients
Akatsi Premier Communications Centre	private individuals	public organisations	small businesses	<25	<25
Zutako Communications Centre	private individuals	small businesses	n.a.	n.a.	<25
Zutako Communications & Business	private individuals	n.a.	n.a.	<25	<25
West Falia Communications Centre	private individuals	public organisations	small businesses	50<75	25<50
Benak Trading and Communications Centre	private individuals	public organisations	small businesses	<25	<25

Table 4-7: Ranking customers by type (in order of importance)(Data source: TAP 1999 and telecom-munication centre expert interviews)

---

17   Those clients who use the services within a certain time period; e.g. daily, weekly, monthly are considered as regular.

Table 4-7 shows the results of a ranking exercise concerning the types of customers the various centres serve and shows that all interviewees regard private individuals as their main customers, followed by representatives of public organisations and entrepreneurs. The respondents also indicate that the approximate proportion of business-related calls is less then 25% with the exception of *West Falia* which claims that between 50 and 75% of calls are business-related. Beyond that, the PCO managers stated that they would consider the majority of clients as non-regular, meaning that they do not use their service so frequently that they got to know the clients.

As far as daily, weekly or seasonal variations in demand are concerned, the results of the survey (cf. Table 4-8) results in the conclusion that there does not seem to be a particular seasonal peak during the year, whereas the weekly demand seems to be highest at the beginning of the week as well as on market days. Consistent answers could also be obtained to the question about the busiest time of day in terms of telecommunications demand: the interviewees all mentioned the (early) morning as the busiest period of the day. Three centres seem to enjoy a further peak in the late afternoon and evenings (cf. Table 4-8).

Name of enterprise	Peak Season		Peak week day		Busiest time of the day	
	First	Second	First	Second		2
Akatsi Premier Communications Centre	December	January	Friday	Monday	8.00-10.30 am	4.00-7.00 pm
Zutako Communications Centre	n.a.	n.a.	n.a.	n.a.	7.00-12.00 am	n.a.
Zutako Communications & Business	August	September	Market day (alternating)	Monday	7.00-9.00 am	6.00-8.00 pm
West Falla Communications Centre	n.a.	n.a.	Market day (alternating)	Monday	6.30-10.00 am	5.30-8.30 pm
Benak Trading and Communications Centre	n.a.	n.a.	Market day (alternating)	n.a.	7.00-12.00 am	n.a.

Table 4-8: Variation in telecommunication service demand (Data source: TAP 1999 and telecommunication centre expert interviews)

101

Finally, the managers of the communication centres were asked to estimate the proportion of clients that come from outside Akatsi Town[18] and to indicate from which settlements these clients originate.

Two of the five respondents could not indicate any concrete figure in response to the first question, whereas the remaining respondents estimated that the proportion of customers from outside town was less than 10%. The answers to the second question are shown in Table 4-9 (also cf. the maps shown in Figure 4-2 and Figure 4-3). Obviously, some answers are misleading as clients from Lomé, for instance, surely pass by the major road and use the facility. For the other communities, except Aflao, it seems plausible that Akatsi is the closest and only possibility to access and use telecommunication facilities.

	Proportion (%) of regular clients from outside Akatsi	Origin of clients (in order of importance)		
		1	2	3
Akatsi Premier Communications Centre	5<10	Anyako	Abor	Tadzevu
Zutako Communications Centre	n.a.	Abor	n.a.	n.a.
Zutako Communications & Business	n.a.	Abor	Gadzepo	Ayitikope
West Falia Communications Centre	5<10	Lome	Denu	Aflao
Benak Trading and Communications Centre	<5	Abor	Wute	Avernompene

Table 4-9: Proportion of regular clients coming from outside Akatsi Town and origin of those clients

4.3.2 The Demand for Telecommunication Services in Akatsi

The last section provided an overview of the supply of telecommunication services in the survey area. The demand for such services will now be the focus. To do so, it is important to determine the proportion of rural households that actually use the services. The link between socio-economic factors and the use of the ser-

---

18   Akatsi town in this sense includes quarters of the town that are not directly located around the centre of town but still belong to the Town Council Area.

vices will also be identified. Service utilisation will be regarded as whether or not the households use the services and how often such use is actually taking place. The demand analysis also aims to apply a methodology that synoptically relates the socio-economic household attributes to service utilisation. This will be achieved by using binomial probit and censored regression models: the former explaining whether people access publicly available telecommunication services, the latter aiming at how the various household characteristics impact on the intensity of telephone usage.

Prior to the multivariate analysis we will, however, show how often and where the households use what kind of facilities and analyse the cross-section of the last telephone calls made and received. Using this method, some key characteristics of individual calls, such as their duration, are captured.

4.3.2.1 Use of Telecommunication Services and the Households' Socio-economic Attributes

In order to obtain appropriate information about the households, the interviews were carried out with the heads of the households. In practice, however, persons other than those officially recorded on the household list might be more actively involved in the economic activities and the management of the household. The survey targeted these individuals. Table 4-10 pays tribute to this fact and shows the actual relationship of the respondents to the official household head[19] for both the screening exercise and the household survey. For the remainder of the analyses, the respondents are considered to be the heads of the households.

Whereas the assessment of household characteristics can be carried out by asking the household representative, telecommunications use is a rather individual activity. It was assumed that the head of the household as defined above would mainly be the person that utilises the telephone. The intra-household penetration of telecommunications use (cf. Figure 4-4) seems to justify this assumption: for all households that use telecommunication services, 82% of the respondents claimed to be the only user with another 14% pointing out that his or her spouse also makes use of telecommunication facilities. Another 4% indicated that three or more members of the household use the services.

On that note, the telecommunications utilisation of the household heads will be the focus of the analyses.

---

19   As mentioned in the sampling frame.

	Screening – survey		Household – survey	
	Frequency	Percentage	Frequency	Percentage
head itself	262	63.9	117	68.8
wife/hus-band	29	7.1	11	6.5
others	119	29.0	42	24.8
Total	410	100.0	170	100.0

	Sex of respondent to screening-survey		Sex of respondent to household-survey	
	Frequency	Percentage	Frequency	Percentage
Male	232	56.6	108	63.5
Female	178	43.4	62	36.5
Total	410	100.0	170	100.0

Table 4-10: Relationship of interviewees to the official household head and sex of inter-viewee

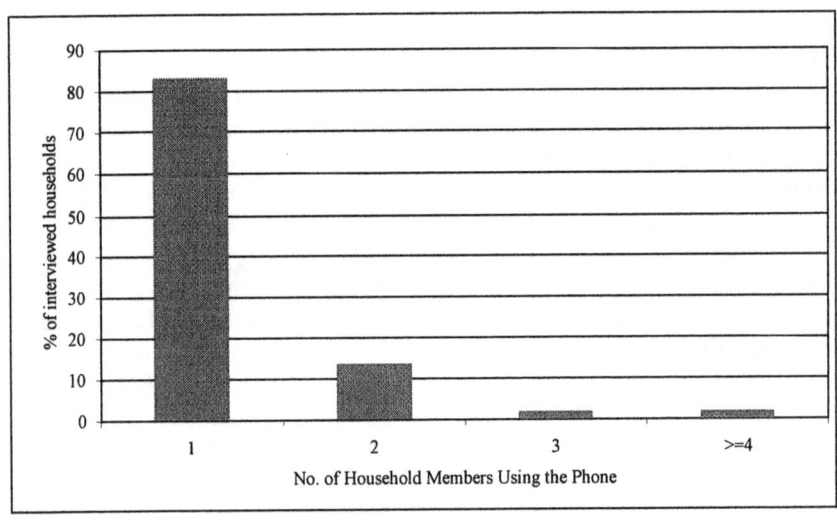

Figure 4-4: Intra-household penetration of telecommunication service use

Proportion of Users and Non-users

Before this analysis will is carried out, it is crucial to define usage. Within the context of this study, telecommunication use is regarded as the *regular* utilisation of any form of voice or fax transmission. Regular use is distinct from occasional use by the users themselves. When asked, how often per specific time unit they would use the telecommunication service[20], 95.2% claimed to use it regularly on an annual basis at least. The other 4.8% claimed to have used the services less than that. Within the remaining part of this analysis, these households will be treated as non-using households. On this basis, the screening survey's result, indicating that 48% of all interviewed households use telecommunications (cf. Table 4-11), should representatively reflect the overall proportion of households that use the services within the communities.

Here, one needs to remember the survey design. It was pointed out in Section 4.1 that the second round of the survey over-sampled using households (also cf. Figure 4-1). Thus, a weighting factor needs to be included (cf. BÜHL, ZÖFEL 1996; Annex 4-3). This weighting will, however, be restricted to the graphic production of the results generated from the univariate and bivariate analyses. The dependency measures and goodness-of-fit values will be produced for both, the weighted and the unweighted data[21] although we will see that the differences between them are minimal.

	Screening –survey			Household-survey		
	Use	Non-use	All	Use	Non-use	All
N	197	213	410	120	50	170
%	48.0	52.0	100.0	70.6	29.4	100.0

Table 4-11: *Ratio of households in both surveys in which the respondents claimed that at least one member has used telecommunication services*

Household Characteristics and Telecommunications Use

In a first analytical step, using and non-using households according to their major socio-economic attributes will be compared. This comparison is based on the hypothesis that the use of telecommunications is determined by human and financial resource factors as listed in Table 2-1. Accordingly, they can be grouped into:
- individual characteristics of the household head,
- factors that refer to the household as an economic unit,
- the community in which the household is actually situated.

---

20   Due to the fact that only five respondents within the household survey actually send and receive facsimiles, voice telephony will be referred to as the *killer application*.
21   For the modeling exercise, the weighed measures are used because the weights were not created for the model.

Related to the individual dimension of the household head, these variables are *SEX, AGE* and his or her educational background (*EDUC*)(cf. Table 4-12).

The household characteristics that will play a core role in the analyses are the main source of household income and the economic status of the household represented by two alternatives. The first is a monetary value reflected by the household's expenditure, the second is a discrete variable representing the economic status of the household within the community.

Variable Name	Description	Variable Scale	Measures
Individual characteristics of household head			
SEX (A5B)	Sex of hh-head	nominal	male/female
AGE (B5C)	Age of hh-head	scale	in years
EDUC	Educational level achieved by hh-head	nominal	illiterate/primary, secondary/post secondary
Characteristics of household			
INCSOURCE	Main source of the household's income	nominal	agriculture and rural industry/trade, services, construction, artisan/ government and administration/others
LNHHINC	Household's expenditure during the last month	scale	in Cedis
INCSTAT	Economic status within community (self assessment)	nominal	poorer than average/about average/richer than average
NOHHMEM	*Number of household members*	*metric*	*number of persons living in the household as defined in Chapter 1*
PART_HH	*Index of membership in community based organisations*	*metric*	*values of 1-4 indicating an increasing degree of participation*
Location of household			
VID	Village of household	nominal	Akatsi/Agbedrafor/Gefia

Table 4-12: Tabular overview of the socio-economic and demographic household characteristics (variables in italics will exclusively be examined in Sections 4.3.3 onwards)

Additional variables that relate to the household as a whole – but which will only play a role in the estimation exercise – are the number of household members (*NOHHMEM*) and the degree of participation of the household within community-based organisations (*PART_HH*) (cf. Table 4-12 and Section 4.3.3). For the analysis, cultural factors and natural conditions will be omitted. This is based on the assumption that – within the limited range of the survey area – homogeneity in terms of both characteristics is given (cf. Sections 4.2.1 and 4.2.2).

Gender

The sex of the household head is believed to have an impact on the decision as to whether telecommunication services are used or not. It is hypothesised that households with a female head are less likely to use the service than those headed by a male. Despite the image that Ghanaian women are heavily involved in business and trade, it is assumed that women are disadvantaged when it comes to telecommunication usage. Other surveys showed that, "in general, women tend to use PCOs significantly less often than men in most developing countries [...]. This reflects socio-cultural restrictions as well as the lower average level of education and employment common among women in most countries" (SAUNDERS ET AL. 1994. p. 248).

The optical impression provided by Figure 4-5 indicates a higher proportion of non-using households being headed by women. This impression is underlined by the low p-value that could be generated by the chi square test. According to these results, our hypothesis seems to be verified.

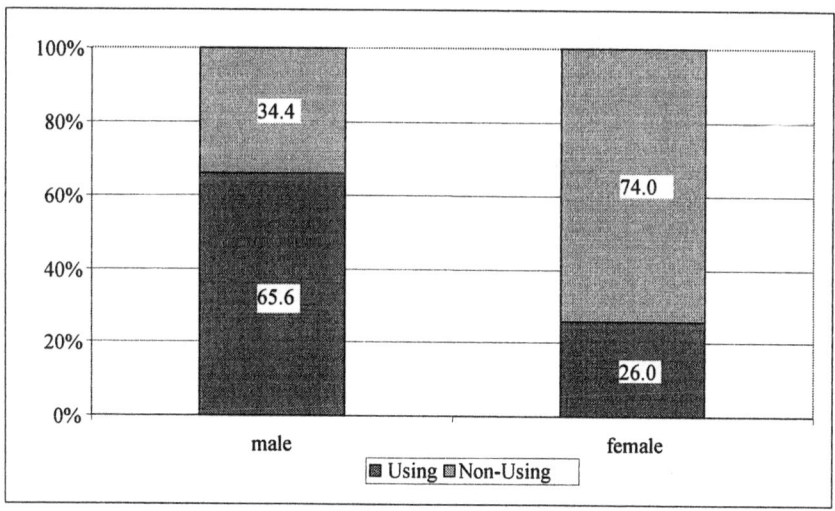

Figure 4-5: Use of telecommunication services and sex of household heads (contingency coefficients weighted/unweighted: 0.37/0.35 at p<0.01)

Age

Similarly, we expect the heads of using households to be younger than those of non-using ones. This is based on the assumption that the older population is more likely to stick to its traditional means of communication. This assumption might be particularly valid since the phone was only introduced 3 years before the survey took place. Younger people will be more open to use newly introduced communication facilities and usually be more active in social and economic terms. A t-test was applied comparing the means for using and non-using households. At a significance level of $p<0.01$, assuming equal variances[22], the alternative hypothesis for both groups having different means can be accepted. The lower mean value for the using household heads furthermore allows the acceptance of the alternative hypothesis (cf. Table 4-13).

Age	**Weighted**		**Unweighted**	
	using	**non-using**	**using**	**non-using**
Mean	42.4	52.2	42.4	52.2
	(13.5)	(17.3)	(13.4)	(17.4)
Sig.	0.001		0.003	

Table 4-13: Use of telecommunication services and age of the household heads (standard deviation in parentheses)

Education

It is a widely accepted fact that there is a general link between openness towards innovations and a higher level of education (cf. Section 2.2). On that note, the level of education achieved by the household head[23] was tabulated against use and non-use respectively.

Highly significant chi² values and symmetric contingency measures were computed for this tabulation. Figure 4-6 shows the positive relationship between these factors and graphically underlines the results from the calculation.

---

22   Supported by the Levene's Test for the Equality of Variances.
23   The respondents that claimed to be educated informally were added to those that enjoyed education up to the primary level.

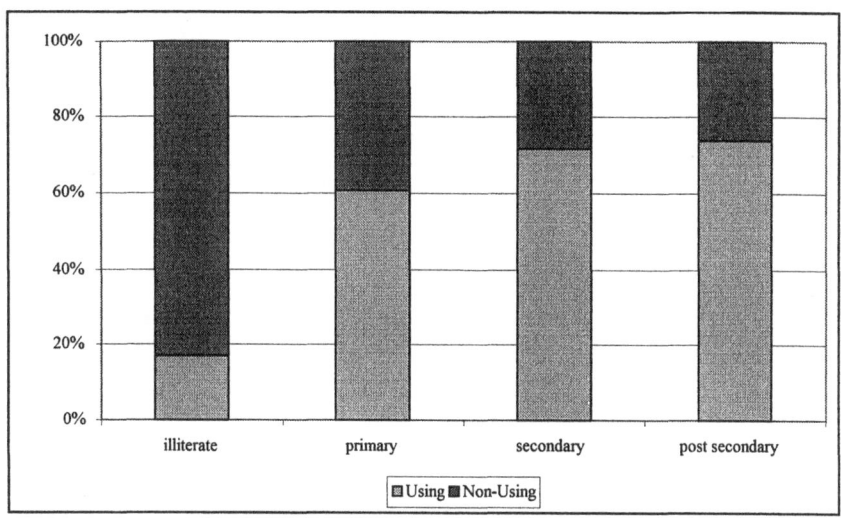

Figure 4-6: Use of telecommunication services and educational attainment of household heads (contingency coefficients weighted/unweighted: 0.42/0.42 at p<0.01)

Income Generation and Economic Status

The first variable that reflects a factor of the household as a whole is the economically most important source of income of the household. In most cases, this would also be the occupation of the household head. The cross-tabulation produced in Figure 4.7 shows the different user rates between the various income source groups. Whereas the lower user rate amongst farmers and people involved in rural industries as compared to traders, etc. might derive from the prevalence of subsistence-oriented farming, the high user ratio amongst public servants cannot be explained that easily. We may draw the conclusion that the government employees might have easy and cheap access to phones at work. As there is a very limited number of public institutions that actually have their own connection to the telecommunication network, this idea can be dismissed. It might be possible, though, that the fact that the Ghanaian government tends to assign public servants throughout the country irrespective of ethnic background, leads to these figures. In this case one would assume a link between work-driven migration and a higher need for communication. Section 4.3.3.2 will elaborate more on this assumption.

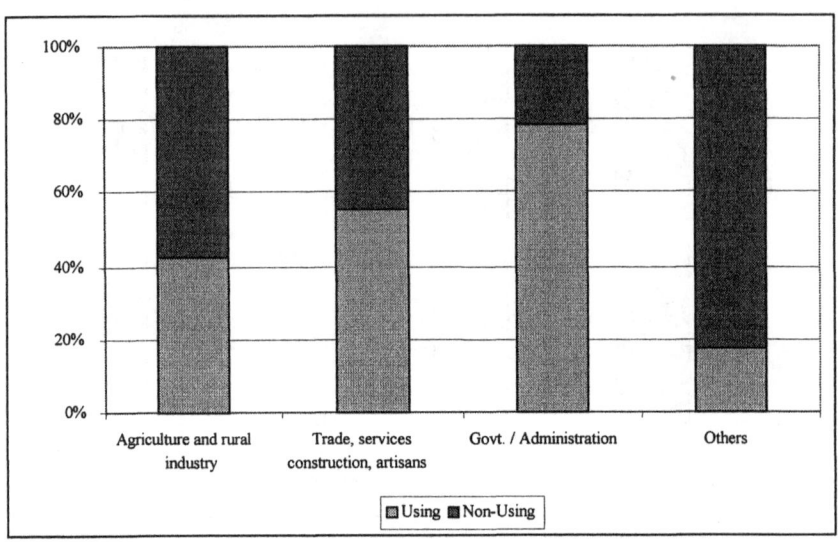

Figure 4-7: Use of telecommunication services and main source of household income (contingency coefficients weighted/unweighted: 0.30/0.27 at p<0.01)

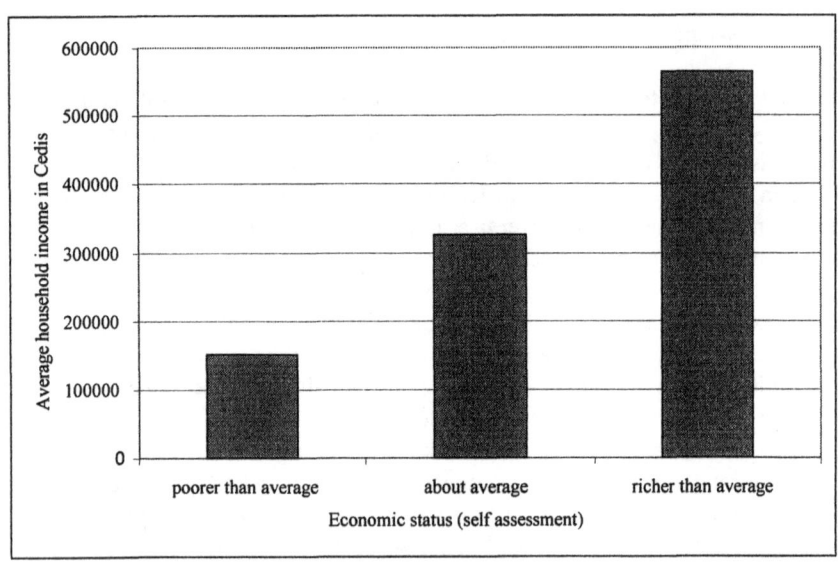

Figure 4-8: Means of the households' monetary income against the economic status that was assessed by the household heads themselves

110

The households' income and wealth level is represented by two measures. First, the logarithmic form of the households' expenditure during the last month is regarded as a proxy for the households' monetary income (*LNHHINC*). Second, a discrete variable reflects the self-assessed economic status within the community. Households consider themselves within the community according to one of three values: *poorer than average, about average,* and *richer than the average* (cf. WEINBERGER 2000). This kind of assessment pays tribute to the fact that the wealth of a rural household may not be appropriately expressed in monetary income but also includes other measures, i.e., the amount of arable land or cattle owned.

Whereas the continuous variable better enables the calculation of marginal effects and uses tangible values, the self-assessment better reflects the deprivation of the household within the community. In light of this argument, both options will be considered and compared, although Figure 4-8 demonstrates consequently high consistency between the two variables.

The dependency between the monetary income[24] variable and whether the households use telecommunications was tested using the t-test for the equality of means. Assuming equal variances, both, weighted and unweighted data result in p-values that allow the clear acceptance of the alternative hypothesis (cf. Table 4-14).

LNINC	Weighted		Unweighted	
	using	non-using	using	non-using
Mean	12.63	11.90	12.63	11.90
	(0.71)	(0.74)	(0.71)	(0.75)
Sig.	0.000		0.000	

Table 4-14: Use of telecommunication services and logged household income (standard deviation in parentheses)

Bearing in mind that the overall income levels between the communities themselves could differ, the respondents were asked to classify the wealth of their household in relation to the other families in their community. Figure 4-9 shows the parallels to the former method of measurement and allows the conclusion that the households' wealth, relative to the other community members, also seems to be positively linked with the use of the phone service.

---

24   The logged value is being used here.

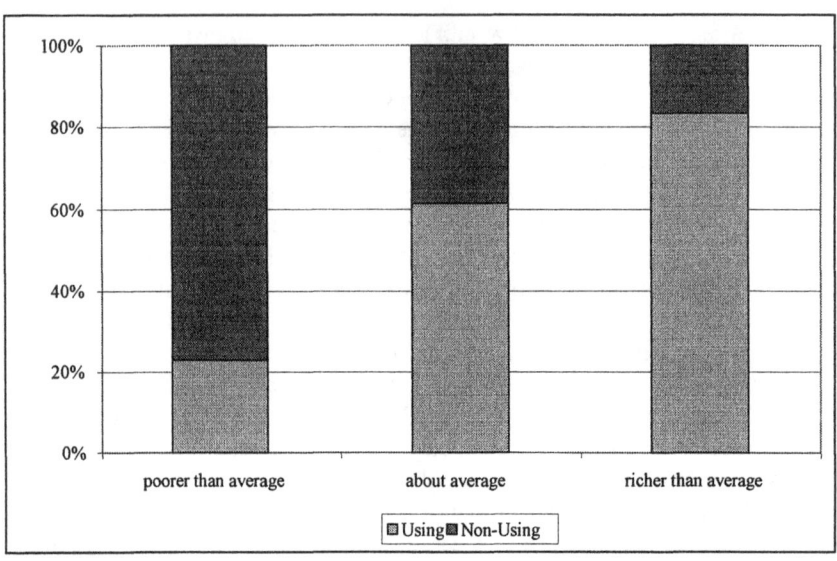

Figure 4-9: Use of telecommunication services and self-assessed wealth ranking within the community (contingency coefficients weighted/unweighted: 0.38/0.36 at p<0.01)

Location

One could consider the variable, representing the community in which the respective household is allocated, as a proxy for whether the community has access to the service or not. On that note we are expecting less households in Agbedrafor and Gefia to use telecommunications than in Akatsi. In view of the higher time and transportation costs that need to be accrued to reach the communication facility this should particularly be the case for Gefia. As expected, the results of the cross-tabulation between this variable and the telecommunication utilisation show that using households in Gefia only account for 18% as compared to around 46% in Agbedrafor and approx. 67% in Akatsi (Figure 4-10).

In terms of the adaptation of the use of the service, Figure 4-11 indicates that – within the first year of introduction, the increase of users in Akatsi was relatively higher than in the other two communities. After 1997, the increase flattens and describes parallel courses with the Agbedrafor figures. Starting on a very low basis, the percentage of households that use telecommunications in Gefia also rises but flattens again after 1998.

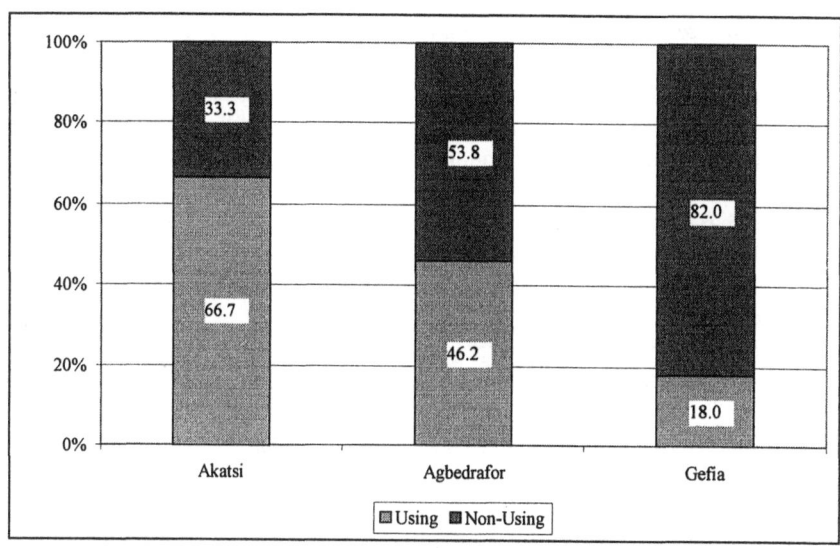

Figure 4-10: Penetration of using and non-using households of telecommunications within the different communities (contingency coefficients weighted/ unweighted: 0.38/038 at p<0,01)

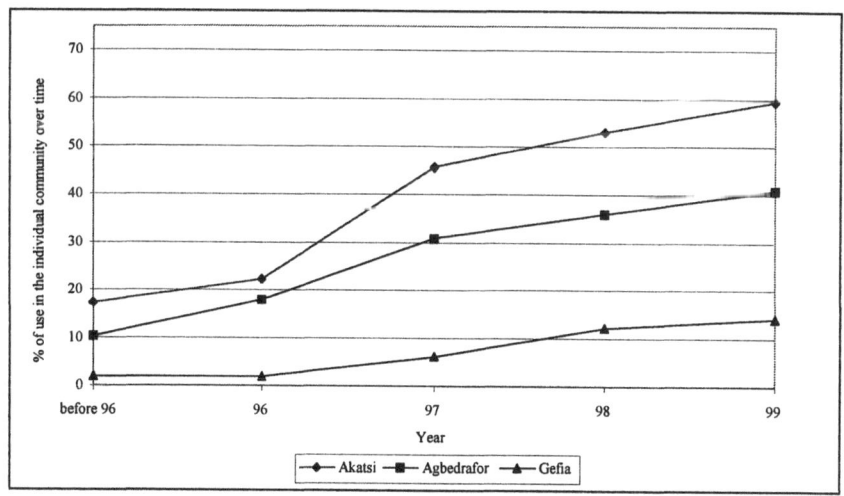

Figure 4-11: Development of telecommunication service utilisation within the communities over time

113

4.3.2.2 Characteristics of Telecommunication Service Utilisation

The previous section showed differences between using and non-using households in terms of their socio-economic attributes. We will now concentrate on some characteristics of telecommunication service use. For this purpose one needs to show
- when the respective household members started to use the services and whether they maintain a list with telephone numbers,
- whether it is justified to assume that use exclusively takes place in Akatsi Town,
- which facilities are used most frequently,
- how frequent the phone services are used for both making and receiving calls, and
- how much money the households spend on utilisation.

Moreover, a cross-section of the last calls that were made and received by the respondents will be looked at and the average duration of and the expenditure on these particular calls will be determined. In this context one should also ask for the quality of the services taking the amount of attempts to complete the outgoing calls as an indicator.

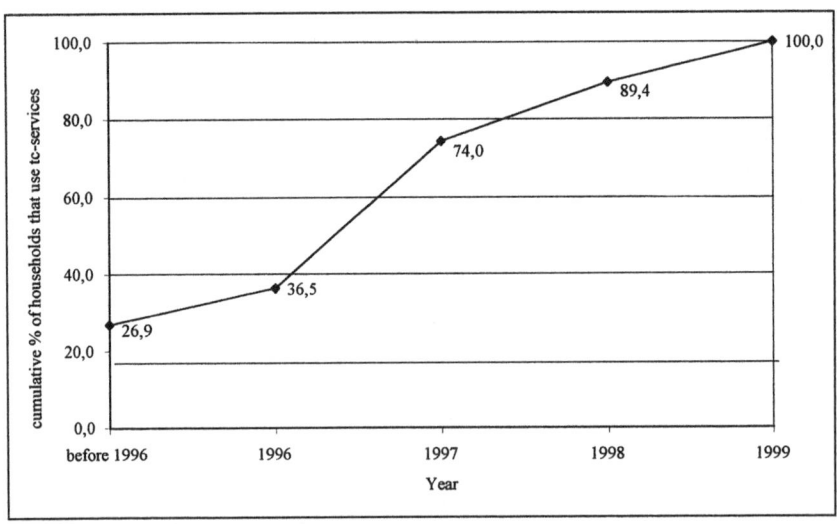

Figure 4-12: Start of service utilisation by using households

114

It became obvious from Figure 4-11 that some respondents already used tele-communication facilities before they were actually introduced in Akatsi. This group accounts for a little more than a quarter of all users (cf. Figure 4-12). The increase in people that started using the services but had never used it before was most significant during the year 1997, i.e. during the first full calendar year of service provision.

Whereas those people using the services before 1996 must have done so in other towns or cities, utilisation now mainly takes place in Akatsi. In fact, all in-terviewees claimed to use telephone and fax services in Akatsi Town itself.

PCOs are by far the most frequented facilities which enable access to tele-phone services. More than 90% of all respondents use PCOs to make their calls: 77% of these exclusively use this kind of access, an additional 15% use PCOs and the available phone booth. The remaining 8% either ask to use the telephones of neighbours and friends or have their own phone either at home or at work. A further 3% of respective users claimed to exclusively make exclusive use of the one public card phone available (cf. Figure 4-13). This seems to be surprisingly low in light of the cheaper tariffs that are required to be able to use the card phone facility. The reasons for that preference are – according to some inter-viewees and the key informants – that the initial investment into a phone card was considered to be too high.

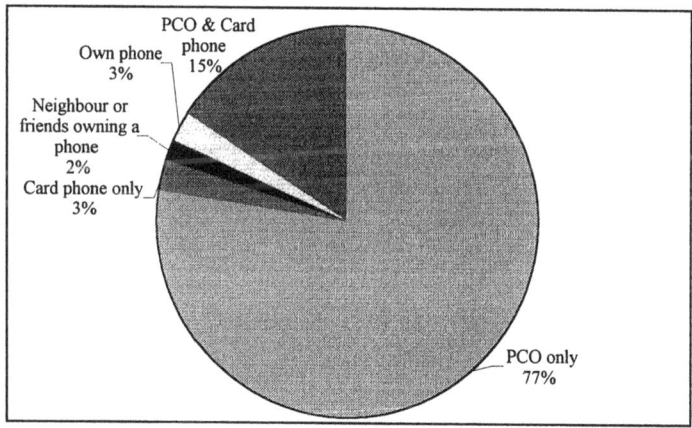

Figure 4-13: Preferred points of access to telecommunication services

This argument becomes more important if one considers the chronic shortage in supply of those phone cards that are relatively cheaper – due to the smaller unit charge. Another aspect is certainly the fact that the public card phone does not provide a reliable service. According to numerous interviewees it is often cut off and the time span until it is fixed again can extend to weeks. Moreover, communication centres accept incoming calls, take messages or fetch people from their houses to answer calls. This may in many cases create a certain loyalty among the customers towards their service providers (AGBEDRAFOR 1999; AKATSI DISTRICT ASSEMBLY 1999).

Intensity and Cost

Figure 4-14 underlines the importance of incoming calls. The graph reflects the annual frequency of outgoing and incoming calls of the households[25]. Based on the fact that 74.2% (N=89) of all using households also receive calls, the importance of incoming calls becomes transparent: 21.3% of those households receiving calls do so less than 20 times per year. Nearly 60% of this group receive between 20 and 200 calls and approx. another 20% is called more than 200 times annually, i.e. more than 4 times per week.

Broken down to different time levels, the average yearly frequency of outgoing and incoming calls lies at 136 and 134 respectively. This comes to a calculated 11 calls per household per month (cf. Annex 4-4).

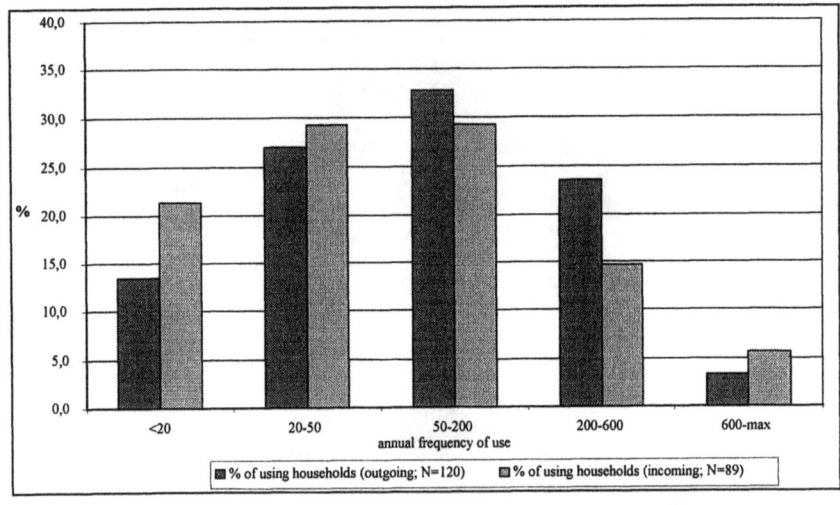

Figure 4-14: Annual frequency of outgoing and incoming calls

---

25 The values are based on the data obtained from the question as to how often household members used the telecommunications facility during the last month.

Another indicator that reflect the degree of variation in people called and the degree of sophistication in use might be whether the household keeps a list with telephone numbers to call and how many records this list contains. Around 78% of all households do actually use and maintain such a list, containing an average of around 10 records.

Expenditure

The indicator that will also be used within the synoptic analysis of determinants of telephone utilisation is the households' expenditure on this use. Table 4-15 is based on the data derived from asking the interviewees how much money their household spent on the services during the last month. Bearing in mind Figure 4-13, it does not seem surprising that some households do use telecommunication services but do not pay for it. On the other hand, the one and only outlier that was identified but not excluded from the data-set does influence the results of the calculation and leads to a rather high standard deviation. The average monthly expenditure of ¢21882.5 was also compared to the overall expenditure of the household. This comparison leads to the conclusion that, within the given sample, telephone services account for approximately 5% of the households' overall expenditure.

	N	Minimum	Maximum	Mean	Std. Deviation
Monthly expenditure on telecommuni-cation services of household (in Cedis)	120	0	400000	21882.50	41299.70
Expenditure on telecommunication services in percent of overall expenditure	120	0	25.5%	5.1%	4.5

Table 4-15: The households' expenditure on telecommunication services

Characteristics of Individual Calls

So far, rather general results related to the intensity of telecommunication service use have been presented. We will shortly sum-up the characteristics of individual calls focusing on their duration and how much they cost. The data that could be obtained by questioning the household heads about the duration of and expenditure on their last outgoing and incoming call is summarised in 4-16. It might be worthwhile pointing out the – on average – slightly longer incoming calls as well as the fact that users are charged around ¢1.000 for receiving calls. Although this is only about a third of what they are charged for making calls, it represents a significant disadvantage compared to owners of residential phones who can receive calls free of charge.

Additional disadvantages become apparent if one considers the average number of nearly 3 attempts to complete an outgoing call. This clearly indicates the severe quality problem of the infrastructure and the costs attached to this problems for both the communication centres and customers.

	N	Minimum	Maximum	Mean	Std. Deviation
Average duration of last call					
Outgoing (in minutes)	109	1	30	5.6	4.4
Incoming (in minutes)	82	1	30	6.3	5.3
Average expenditure on last call					
Outgoing (in Cedis)	108	0	19000	3020.6	2517.2
Incoming (in Cedis)	84	0	5200	1002.4	1145.1
Amount of trials to complete call (referring to last outgoing call)	99	1	21	2.7	2.26

Table 4-16: Major characteristics of incoming and outgoing calls

4.3.3 Determinants of Telecommunications Utilisation

In this section we go beyond univariate and bivariate analyses in order to consider the factors that determine the utilisation of telecommunication services and the intensity of this utilisation from a synoptical perspective. To do so, two steps are required.

The first will deal with determining why people use publicly available telecommunication services: previous sections showed that linkages can be drawn between the various household attributes and telephone use. It became obvious, for instance, that the ratio of using and non-using households is different within the three communities. In this respect, multivariate methods may help to show the contribution of the community variable – reflecting distance to the services – to the decision of the household head to use the facilities relative to the other household attributes mentioned (also cf. SAUNDERS ET AL. 1994; TORERO 2000).

The second step is a complementary one. It aims at understanding how the household characteristics impact on the intensity of telephone usage, which is reflected by the households' expenditure on the services available[26]. One might, for instance, be able to show the effect the household income has on the intensity of telephone utilisation.

4.3.3.1 Motivation and Background of Estimating Use and Intensity of Telecommunication Services

To carry out both steps, regression analysis will be utilised. This tool enables us to identify the direction, magnitude, and significance of the parameters influencing telecommunication use (cf. CHATERJEE, PRICE 1995; GUJARATI 1995).

To achieve this, an approach developed by MITCHELL (1978) and modified by TORERO (2000) will be followed. Both authors measure the demand for residential telephones. As this work aims to estimate the access to and use of public facilities, the concepts proposed by the authors need to be modified. Although different in terms of their conceptual background, the theoretical starting point of the estimation exercise remains the same. From consumer theory a utility function can be set up (cf. VARIAN 1995). This function contains the indirect utility (V) of telecommunication services (tc) and other goods (x). It can be expressed as:

$$U(x, tc) = x + V(tc) \qquad \text{(Eq. 1).}$$

The user of the service will maximize utility according to the budget constraint he or she faces. For residential phones, MITCHELL (1978) and TORERO (2000) proceed in adding to x – representing the bundle of other goods – the product of the parameter of using the service and the monthly subscription fee. On top of that, a vector of prices paid for the calls is added (cf. Annex 4-5).

For the estimation of use of public telephones the following changes have to be undertaken. First, the access costs need to be measured differently: the minimum fixed costs a user accrues to make a call is the monetary expression of his or her time and money spent to reach the phone, making the call, and going back again. On top of that, he or she needs to pay at least a minimum price to establish a phone call, usually the minimum price for one unit.

At this stage one needs to consider one important thought: MITCHELL (1978) adds the fixed amount of telephone rental to the consumer surplus. This is due to

---

26   In this context, no distinction will be made between local, trunk, and international calls.

the possible scenario that the telephone owner receives calls without making any. Remember that the users of PCOs – at least in our case – are willing to pay more for their call in view of the fact that they can be called and messages can be left with the PCO managers. Moreover, our way of calculating the minimum fixed cost a user will have to pay in order to make a call can easily be modified considering the possibility of receiving a phone call. In that case, the minimum unit price to make a call is just replaced by the flat rate demanded if one receives a phone call at the PCO (cf. Sections 4.3.1.3 and 4.3.2.2).

In order to maximise the utility, one needs to consider the households' budget constraint expressed as:

$$y = x + \partial(t + m + p * tc) \qquad \text{(Eq. 2).}$$

With y being the income expressed in units of x, and t reflecting the time to reach the phone; m stands for the cost to establish the call or the flat rate for receiving calls respectively. The quantity of units used (tc) times the vector of tariffs for local, trunk, and international calls (p) reflects the overall expenditure on telecommunication services. The parameter ∂ describes if the consumer uses the telephone services at all ($\partial = 1$) or if he or she does not ($\partial = 0$).

To be able to estimate whether or not the respective household members will use the service, it is crucial to calculate the difference between the utility of using the service and the cost of this utilisation. In terms of the decision whether to eventually use the service, this means that the consumer surplus from accessing the phone ($V(tc)$) should be greater than the cost of going to the phone and using it:

$$[V(tc)] - [\partial(t + m + p * tc)] \geq 0 \qquad \text{(Eq. 3).}$$

Reconsidering the optimisation problem in this context leads to the maximisation of the difference mentioned and is in line with MITCHELL (1978) and TORERO (2000):

$$\underset{\partial, tc}{Max}[V(tc) - \partial(t + m + p * tc)] \qquad \text{(Eq. 4).}$$

As we are for now concentrating on access to the phone, hence, the time to reach the phone and the money to initiate a call, Equation 3 leads to the statement that:

$$\partial = 1 \text{ if } R(p) \geq t + m$$
$$\partial = 1 \text{ if } R(p) < t + m$$

and   (Eq. 5).

With $R(p)$ being the consumer surplus or reservation price respectively (cf. VARIAN 1995). This value is given as:

$$R(p) = \max_{tc}[V(tc) - p * tc]$$ (Eq. 6).

How these thoughts will be approached econometrically, will be explained in 4.3.3.3.

In order to estimate the *intensity* of service use, these considerations need to be modified. The motivation behind doing so stems from the fact that the household characteristics may contribute differently to the decision by household individuals to use the services and the amount they are willing to spend on use.

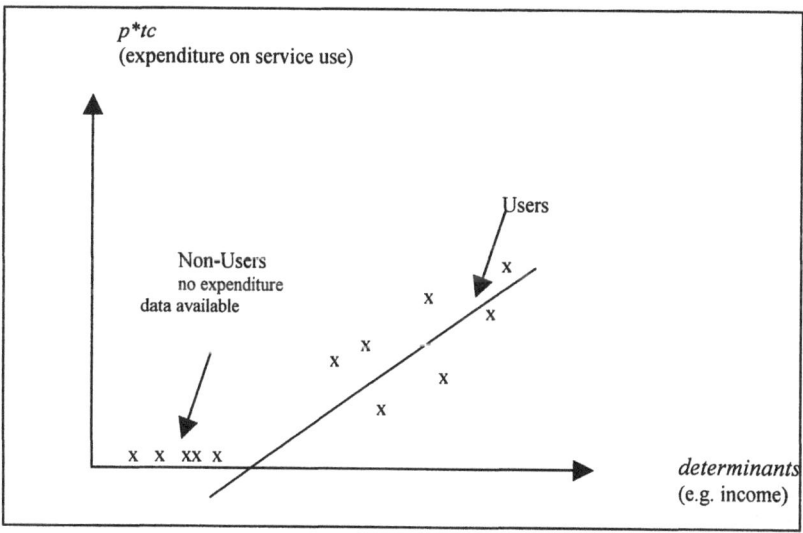

Figure 4-15: Graphic illustration of tc (cf. GUJARATI 1995)

For instance, the probit estimation may indicate a negative impact on the propensity of using telecommunications if the respondents are illiterate. This difference may, however, vanish if the intensity is estimated. This might be due to the limited variation of money spent on phone use between the users that enjoyed different levels of education. This fictional example also points at another problem: As the variable that represents the intensity of use will be the monthly expenditure on telecommunication use by a household ($p*tc$), the problem occurs that the non-users are excluded from the considerations.

To tackle this, an extension of the probit analysis – namely the tobit limited dependent variable model – will be applied. This form of regression analysis enables the estimation of censored variables and therefore allows us to include all households into the estimation, whether they are users or non-users of telecommunications (cf. Figure 4-15).

From a conceptual view, Equation 3 needs to be reconsidered and one needs to assume that the utility is equal to the costs and, hence the entire expression is equal to zero. Assuming that m can be omitted because its value is negligible, $p*tc$ can be estimated as follows:

$$p * tc = V(tc) - t \text{ if } \partial = 1$$

(Eq. 7).

$$p * tc = 0 \quad \text{otherwise.}$$

The econometric solution of the tobit estimation will be presented in 4.3.3.4.

4.3.3.2 Specifying the Estimation

Before the results of the modelling exercise can be produced, the empirical set-up needs to be explained. This will mainly be done by taking the dimensions of ICT use into account and elaborating on the variable set-up of those dimensions (cf. Chapter 4.3.2.1). Against the background of the slightly different concepts of the models, this will be done separately, relating the variables to the hypotheses that will be tested by the regression analyses (cf. Figure 4-16).

Estimating Use

To estimate whether the households use the services or not, the hypotheses discussed in Section 4.3.2.1 are reconsidered and unfold as follows:

A negative impact of the household being headed by a female person on the propensity of telecommunication services is hypothesised.

Similarly the increase in age of the respondent is expected to reduce the propensity of telecommunications use.

Thirdly, with an increase in the level of education, the respondents are said to more likely use telecommunication services. As education is represented by a categorical variable, it needs to be split up into dummy variables (cf. GUJARATI 1995): With not being formally educated representing the reference value, the dummy variables take the value 1 if the respondents did enjoy education up to the primary, the secondary, and postsecondary level respectively (variable names: *EDU_PRIM, EDU_SECO* and *EDU_POST*).

We are furthermore hypothesising a high impact of income measures on the propensity of service use. The household characteristics that relate to the economic status of the households are twofold (cf. Table 4.3.2.1): the proxy variable for household income and the self-classified status of wealth. As these variables are expected to be very important for the service utilisation, they will each be integrated into an alternative estimation set-up. For an increasing metric income measure, a higher propensity of telecommunication use is hypothesised. The dummies representing the self-assessed economic status within the community take the value 1 if the interviewee considers the economic status of his or her household to be below or above the community's average (variable names: *POORER_A, RICHER_A*). All other cases are treated as "being average".

Related to the dummies representing the home community of the households, we expect the propensity of accessing and using telecommunication facilities to decline with the change in communities (variable names: *DUMAGBE, DUMGEF*).

On the basis of these hypotheses, the estimation set-up can be defined as follows[27]:

$$Y_{use} = f(Sex, Age, Education, Income, Location) \qquad \text{(Eq. 8)}.$$

Estimating the Intensity of Use

The idea behind measuring the intensity of telecommunications use is different from the section above. How the defined determinants contribute to the intensity of telecommunications use will now be estimated. Due to this target and the left censored character of the dependent variable, the hypothesis related to the personal and household-related dimensions have to be modified.

On the personal level, being female is believed to have a negative impact on the intensity of phone use reflected by the households' overall expenditure on the services. Thus, the disadvantaged position of women does lead us to the hypothe-

---

27   For the mathematical implementation of the two alternatives that derive from the different income measures care for Annex 4-11.

sis that their households are not only less likely to access telecommunications, but also to use it less intensely than the others.

In relation to the age of the respondent, one also could expect a negative relationship between this variable and the intensity of telecommunications utilisation: even if older people use the phone, their assumed lower economic and social interaction would have a limiting impact on the intensity of telephone use.

It is furthermore hypothesised that the intensity of utilisation increases with the level of education of the household's head. Apart from the links that certainly exist between income level and the activities of a household as well as the education of its head, individuals that achieved a post-secondary level must have done so in a larger town outside Akatsi. This might have increased their social and economic contacts beyond the Akatsi District and eventually lead to a higher intensity of telephone use.

On the household level, consumer theory tells us that higher income results in an increase in consumption and consequently also in telecommunications use. Beyond that hypothesis, one can also state that the usage becomes more intense with the growing involvement in economic activities.

With regards to household-related characteristics that are expected to influence the intensity of service utilisation (cf. Figure 4-16) we expect a lower intensity of telecommunication utilisation if the household is involved in agriculture. This is derived from the knowledge about the high degree of subsistence farming and the resulting limited need of business-related information exchange. A positive interrelationship is expected in the case of government employees. This is not due to the nature of their employment but rather due to the fact that the Ghanaian government tends to appoint their higher-ranking employees irrespective of ethnic background to other regions than where they originate from (cf. 4.3.2.1). Thus, the traditionally strong family and personal ties can be maintained through easy communication and lead to higher expenditures on telecommunication services. A negative relationship between other sources of income and the intensity of use is assumed because this group mainly consists of pensioners and households that do not have a reliable and steady source of monetary income[28].

Second, the number of members in the household needs to be considered (variable name *NOHHMEM*): although the penetration of use within the households is low, it is assumed that an increasing number of household members also increases the intensity of phone use. This is based on the idea that there are more reasons to transmit information the more people live together.

---

28   The dummy variables representing the various sources of income are *INCS_AGR*, *INCS_GOV*, *INCS_OTH* with the households mainly being involved in trade representing the reference.

Moreover, an increasing value of the index representing the involvement of the households in community-based organisations and groups would usually result in an increase in service utilisation (variable name: *PART_HH*). The assumption behind this hypothesis is that those households that are more involved in economic activities will use more sources of information and therefore be more involved in farmers' or traders' organisations and associations.

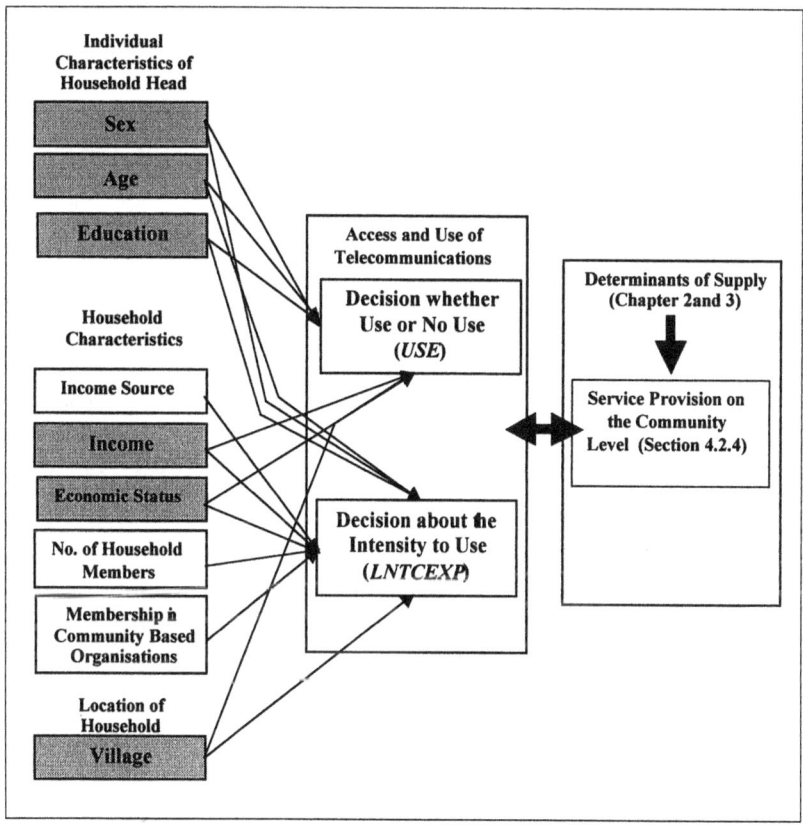

Figure 4-16: The determinants of access to and use of telecommunication services on the level of the individual household. Note that status of employment is only tackled descriptively and is not part of the modelling exercise. Note furthermore the different shading indicating whether the variables are an integral part of both estimations or whether they are only used for the tobit part.

Concerning the dummies representing the households' location, we also expect the intensity of using telecommunication facilities to decline with the change in communities. However, this decline will be much greater if the fact is considered that Gefia's population needs to accrue much higher time and monetary costs to reach the services than that in Agbedrafor and Akatsi Town.

Building upon those assumptions and hypotheses the model can be set-up as[29]:

$$Y_{intensity} = f(Sex, Age, Education, Income, Household Members, \quad \text{(Eq. 9)}.$$
$$Main\ Source\ of\ Income,\ Participation,\ Community)$$

4.3.3.3 Determinants of Telecommunications Use – Empirical Results

Probit

First of all, the binomial probit model requires a stringent definition of the dichotomous variable *use* of telecommunication services by comparing $R(p)$ with the minimum cost of access, $(t + m)$ leading to ∂^*. Bearing in mind that we are aiming for an estimation of the dichotomuos variable containing values of 1 (using) and 0 (non-using) but realising that ∂^* is continuous in its values, leads to the expression that (cf. MITCHELL 1978, TORERO 2000, GUJARATI 1995)[30]:

$$\partial^* = \beta'x + u \qquad \text{(Eq. 10)}.$$

assuming ū is the normally distributed error term with mean zero and variance 1. Given that we only observe whether households use telecommunications or not, the observation will be

$$\partial = 1 \text{ if } \partial^* > 0 \text{ and}$$
$$\qquad\qquad\qquad \text{(Eq. 11)}.$$
$$\partial = 0 \text{ if otherwise}$$

Given symmetry and using the maximum likelihood function (cf. Annex 4-6) the probability of $\partial = 1$ will be:

---

29 For the mathematical implementation of the two alternatives that derive from the different income measures cf. Annex 4-12.

30 A more detailed explanation of the probit regression can be found in Annex 4-6.

$$\text{Pr } ob[\partial^* > 0] = \text{Pr } ob[u < \beta'x]$$
$$= \Phi(\beta'x)$$

(Eq. 12).

Separate from the variables' individual measures of significance, the overall quality of the probit estimation will be examined by applying the following methods.

The first, R^2, is known from linear estimation and was adjusted to the non-linear character of the probit regression model. This so-called Pseudo R^2 can be obtained using different methods. In line with the $STATA^{TM}$ statistical software package that was used, the Aldrich-Nelson-Pseudo R^2 measure will be used[31].

Of greater importance in the framework of the probit estimation is the significance level of the overall model, based on the log-likelihood values and the related chi^2 statistic. For the latter the Wald-Chi^2 value for non-linear regression models is suggested (GUJARATI 1995).

Probably the strongest argument reflecting the quality of the binomial probit model is the production of a table that shows the estimation's prediction accuracy. In the concrete case, this happens by tabulating the observed counts of telecommunications use – and non-use respectively – against the predicted ∂^*. The latter is noted as a 1 (using), if $\partial^* >= 0.5$ and 0 (non-using) if otherwise.

These methods are applied to the two alternative probit estimations that are merely distinguished by different income measures. As apparent from Annex 4-7, the goodness of fit measures produce acceptable results for both alternatives: the Pseudo R^2s take values of above 0.30 and the levels of significance of the overall fit of the estimations are very high (p<0.01). Moreover, the accuracy of prediction is around 78% for both alternatives presented.

On the level of the variables, we will now proceed to test the stated hypothesis by comparing them to the estimation results (cf. Table 4-17; for a summary of these results cf. Annex 4-8).

To begin with, both alternative estimations show a highly significant relationship (p<0.01) between the sex of the respondent and the propensity of the household actually using telecommunication services. Apart from showing the expected negative sign, the marginal effects (-0.24 and -0.26) generated are the highest indicating the importance of this attribute for telecommunications utilisation.

---

31 Some authors suggest different methods of measuring the Pseudo R^2 (cf. MADDALA 1983). However, due to the overall skepticism towards the power of validity of this measure, this measure will not be elaborated further: it will just add information to the overall analysis of the fit of the chosen approach (cf. GUJARATI 1995).

The hypothesis that there is a significant relationship between the age of the respondents and the propensity to use telecommunications cannot be supported as clearly: For the set-up that contains the continuous measure for the household's income, a significance level of 0.058 is computed. For the second alternative, no significance at the 10% level could be achieved. Whereas both coefficients obtain negative values, their marginal effects are relatively low. This result may be generated by the fact that phone services are easy-to-use and do not create much reluctance within the older population.

Telecommunication service use (*Yuse*)	Alternative 1		Alternative 2	
	dF/dx	Robust Std. Err	dF/dx	Robust Std. Err
sex of respondent	-.2424003***	.0873132	-.261563***	.086028
age of respondent	-.005116*	.0026775	-.0039914	.0028877
primary education achieved (edu_prim)	.2050924**	.0976226	.2160738**	.0962428
secondary education achieved (edu_seco)	.1473076	.1003003	.1940599*	.0855625
post secondary education achieved (edu_post)	.1717017	.0894068	.2039191*	.0863314
household income (lnhhinc)	.2087056***	.0631221	n.a.	n.a.
self assessed econ. status richer than average (richer_a)	n.a.	n.a.	.1074597	.1345818
self assessed econ. status poorer than average (poorer_a)	n.a.	n.a.	-.2063459**	.101627
Agbedrafor (dumagbe)	.0612721	.0980511	-.0112843	.0962819
Gefia (dumgef)	-.1552482	.1192659	-.2349069**	.1236071
Pseudo R²		.3304		.3054

*significant at p < 0,1   ** significant at p = 0,05   *** Difference significant at p = 0,01, Pseudo R² = 0.3304

Table 4-17: Results of the probit estimation (for details cf. Annex 4-7)

The models' significance levels also differ if the educational background of the respondents is considered. All parameters do contain positive signs, which is in line with the stated hypothesis. The difference between the produced marginal values indicate that the propensity of telecommunication utilisation does not rise with increasing levels of education. Moreover, the result shows that – compared to the basis (defined as not being formally educated) – having enjoyed primary education has the strongest influence on telecommunication service utilisation (cf. Table 4-17).

The model also supports the hypothesised positive relationship ($dF/dX = 0.21$ at $p<0.01$) between the household's income and the propensity of using telecommunications. The alternative set-up, measuring income in categorical terms, shows that the results for the dummy variable representing those households that claim to be poorer than the community average are significant at $p<0.05$ and obtain relatively high marginal effects ($dF/dX = 0.21$). Consequently, being richer than the average household in the community does not have a significant impact on the propensity of telecommunication use.

The hypothesised impact of the location on the propensity of use is estimated using the dummies for Agbedrafor and Gefia. The former has no significant impact. Concerning the latter, the results differ between the two equations: the model metrically measuring income does not compute a significant coefficient at the 10% level ($p=0.16$). This is not the case for the alternative estimation where $p<0.05$. As far as the direction of the effects is concerned, both alternatives are in line. They support the hypothesis that the sheer distance to Akatsi has a prohibitive impact on telecommunication use. This is particularly valid for the households located in Gefia.

4.3.3.4 Determinants of the Intensity of Service Utilisation

This section discusses the results that were derived from estimating the intensity of use, represented by the households' expenditure on telecommunication services. For this purpose two alternative tobit models could be identified. They contain the same variables as the probit equations but also include the variables from which a particular contribution to the intensity of telecommunication utilisation is expected (cf. Section 4.3.3.2).

Econometrically, the tobit model divides the observations into two groups: one group – in the concrete case the users of telecommunication services – of which information on the independent as well as the dependent variable are available. And another group of which only information about the regressors can be identified. On this basis, a so-called censored variable (z) can be estimated (GUJARATI 1995)[32]:

$$z_i = \beta' x_i + u \text{ if } \qquad \beta' x_i + u_i > 0$$
$$z_i = 0 \text{ otherwise,}$$

(Eq. 13)

---

32 The more detailed explanation of the tobit regression can be found in Annex 4-9.

assuming the residuals being normally distributed and obtaining a common variance.

	Alternative 1		Alternative 2	
Expenditure on the services during the month before the interview (*Ylntcexp*)	Coef.	Std. Err	Coef.	Std. Err
sex of respondent	-2.861864***	.8887017	-3.35302***	.9060806
age of respondent	-.058514**	.0285072	-.0356433	.0293543
primary education achieved (edu_prim)	3.579792***	1.262319	3.652683***	1.291089
secondary education achieved (edu_seco)	2.784674*	1.54874	3.583043**	1.567375
post secondary education achieved (edu_post)	2.993444*	1.647255	3.853434**	1.643787
main source of household income from agric. (incs_agr)	-.4660754	.9794701	-.8048257	1.011193
main source of household income from gov. (incs_gov)	1.193293	1.262673	.400474	1.289544
main source of household income from other (incs_oth)	-1.807188	1.741143	-2.395335	1.773378
household income (lnhhinc)	2.277078***	.6237518	n.a.	n.a.
self assessed econ. status richer than average (richer_a)	n.a.	n.a.	1.989797	1.293691
self assessed econ. status poorer than average (poorer_a)	n.a.	n.a.	-2.272576**	1.00606
number of household members (nohhmem)	.1885548	.1245833	.2338107*	.1258232
degree of participation in community-based organisations (part_hh)	.3917128	.5348763	.2862266	.5451377
Agbedrafor (dumagbe)	.1597895	1.024162	-.7668153	.9864155
Gefia (dumgef)	-2.378048**	1.174898	-3.243002***	1.154301
cons	-22.52781	7.984212	5.460053	2.467585
Goodness of Fit	Number of obs=157 LR chi2(13)=100.03 Prob > chi2=0.00 Log likelihood=-353.60 Pseudo R2=0.1239		Number of obs=157 LR chi2(14)=95.35 Prob > chi2=0.00 Log likelihood=-355.94 Pseudo R2=0.1181	

significant at p < 0,1   ** *significant at p = 0,05*   *** *Difference significant at p = 0,01*

Table4-18: Results of the tobit estimation (for details cf. Annex 4-10)

The overall goodness of fit of tobit estimations is most appropriately measured by the chi^2-based level of significance. Judging from the values presented

(Prob > chi2=0.00) in Table 4-18 both regression alternatives show significant results.

On the level of the household heads' individual characteristics, the results indicate that even if the services are used by the female household heads, the intensity of this use is significantly lower.

Concerning the age of the household heads, the tobit method generates a difference between the two alternative set-ups: with $p<0.05$ the estimation including the numerical income measure does generate a significant result with a relatively low – but in line with the hypothesis – negative coefficient. The other option does not generate a significant result although the parameter's direction and dimension are consistent (Table 4-18).

The results produced by the tobit estimation related to the educational level are highly significant ($p<0.01$) for both alternatives when it comes to primary education. The level of significance varies if secondary and post-secondary education is analysed: the first estimation produces significant results at the 10%, the second at the 5% level. The parameters all obtain positive signs, supporting the hypothesis of a positive relationship between the intensity of telecommunication utilisation and the level of education obtained by the households' heads. Similar to the probit regression, the magnitude of the coefficients indicate that the intensity of telecommunication utilisation does not rise with the level of education: compared to the basis, having primary education has the strongest influence on the regressant (Table 4-18).

Despite carrying the signs expected, the coefficients holding for the main sources of the households' income show no significant difference from zero. The positive relation that was expected between the intensity of telecommunications use and the respondent being employed by the government does not show up empirically here: whereas they are the income group with the highest percentage of users (cf. Figure 4-7), the estimation does not show that public servants use the phone more than other income groups. On a relaxed stand, however, the validity of the assumed negative relationship of the income group containing pensioners and those households that do not have a steady source of monetary income, is open to discussion (cf. Table 4-18).

The hypothesised strong positive relationship between the households' monetary income and its intensity of using telecommunications was supported by the estimation (Coef. = 2.28 at $p<0.01$). The optional estimation set-up shows similar results to the probit estimation. The variable representing the households that claim to be poorer than the community's average is significant at the 5% level. The relationship also carries the expected negative sign with a relatively high coefficient (Coef. = -2.27). Yet again, the parameters standing for being richer than the average household take the expected positive values but fall outside an acceptable level of significance (p=0.13).

As far as the coefficients that relate to the number of household members are concerned, both show positive values of relatively low magnitude. With p=0.13 for the first tobit estimation and p=0.07 for the second, the hypothesised link between increased household size and its overall telephone expenditure can only be accepted with reservation.

The index representing the degree of participation of the household in community-based organisations does not statistically interrelate with the dependent variable for either model. This indicates that the expected link between expenditure on telephone services and involvement in community-based organisations cannot be computed in this way. One reason for this might be that it is not justifiable to conclude that participating households are generally more involved in economic activities than others.

In terms of the households' location, the results of the tobit equation support the great decline in intensity of service utilisation if Gefia is considered: both alternatives come up with a high negative value on a high level of significance (p<0.05 and p<0.01 respectively). This level is not at all acceptable for the parameters related to the dummy representing Agbedrafor (cf. Table 4-18). On this basis, parallels with the probit estimation can be drawn and the hypothesis accepted that in Gefia – besides the lower propensity of use – people also use the telecommunication services in a less intense manner.

4.3.4 Conclusion

This chapter aimed at answering the question of who is actually using telecom services and how this utilisation could be characterised. The descriptive supply and demand analysis clarified the kind of socioeconomic environment in which this happens and showed how supply and demand actually function within this given context. By applying more complex methods, the degree and direction of the effects of household characteristics on telephone use were determined. Looking at the results in a synoptic manner, we can now draw some conclusions from the determinant estimation (cf. Annex 4-8).

First of all, the analyses indicate that the chosen personal and household-related characteristics explain the use of telecommunication services and its intensity. This applies for both estimations and the respective alternative set-ups: Apart from the high goodness of fit, rather consistent signs of direction and significance levels were produced.

As far as the personal dimension of the household head is concerned, gender and formal education appear to have the strongest influence on the propensity and the intensity of telephone use. The monetary income also shows a clear positive impact.

The variables that were specifically integrated into the tobit model due to their expected impact on the intensity of telephone use (income source, participation in community-based organisations, and household size) did not generate the expected results. This holds especially true for the participation by the households in community-based organisations and the main source of household income.

Interestingly, it seems as if both propensity and intensity of service utilisation do not necessarily increase with the degree of wealth or education: Both estimations do not show a significant relationship between the propensity to and intensity of phone use and being wealthier than the average or having enjoyed higher than primary level education. This result would, for the survey area, support the notion that the services are not a tool exclusively used by the rich and educated local elite.

The location of the household does not play the anticipated role. Within the regression exercise, a statistical difference in access and use of the services by people living in Akatsi and Agbedrafor could not be shown. However, the parameters attached to the Gefia dummy allow the acceptance of the alternative hypothesis that physical constraints and resulting prohibitive transportation and/or time costs exclude larger parts of the households from using telecommunications and, if they use it, significantly reduce the intensity of doing so. Hence, it seems to be sufficient to be located in Agbedrafor to have an equal propensity of use and a similar intensity of doing so compared to Akatsi where the services are provided. This result surely supports the definition of universal access chosen as being appropriate in the specific rural environment.

Finally, some problems that might be inherent to the approach chosen need to be discussed. The significance level related to the Gefia dummy was – at least for the probit set-up containing the metric income variable – not as strong as expected. This might reflect the fact that the various factors used are not necessarily uncorrelated to each other. In the concrete case, the limited degree of the impact of the location might be due to its correlation with income. And indeed, the statistically more significant results provided by the option that integrated a categorical way of wealth assessment may support this notion. Moreover, there usually is an interdependence between income and the level of education.

Whereas those problems were addressed in the test for multicollinearity[33] and not found to be prohibitive for the application of the chosen set-up, some degree of violation of the basic assumption of multivariate regression models is inherent in most relationships and can barely be avoided (GUJARATI 1995). These kinds of violations – another example might be the clear two-dimensional relationship between being female and a generally lower level of income – might also be the reason for further patterns of changing significance levels between the different alternative set-ups. The most obvious are that the variable *AGE* in both modelling exercises obtains acceptable significance levels if income is represented by the continuous variable, whereas this relationship can statistically not be accepted for the alternative set-ups. The opposite is the case with the dummy variables reflecting the level of education achieved by the respondents: their measured impacts become statistically significant if the economic status of the household is represented by the categorical variable.

Despite these problems, the results can be regarded as stable and form a basis for extending the work by asking how we can characterise the purpose of and benefits from telephone use. This will be done in the next chapter.

---

33 This was done by cross-checking for a correlation between the individual variables and, in the case of the relationship between income and education which is usually expected, by running the regression without the education dummies.

5 Benefits from Telecommunications Access and Use

So far, this study has concentrated on explaining the character and determination of telecommunication service supply and demand. In this section, the impact of telecommunication services on the community and especially the household level will be assessed.

For this purpose, we need to go beyond just looking at the technical characteristics of service use and consider the information this use is actually transmitting. By doing so, the major purposes of telephone utilisation will be elaborated by asking
- what the interviewees use the services for, and
- where they see the specific benefits from that use.

Additionally, we will ask where key informants and experts see the specific benefits from telephone use for both households and communities. The discussion of those benefits will be structured according to Section 2.3, which referred to the lower transaction costs generated through telecommunications and the benefits from this cost reduction for different actors and organisations in rural society.

Section 5.2 then discusses the benefits of service utilisation on the household level utilising the time-geography-approach of HÄGERSTRAND. On that basis, the benefits of telecommunications use as opposed to its best alternative are measured. Moreover, the increase in the amount of information available and the impact of this increase will conceptually be introduced and underlined by giving examples.

5.1 Purpose and Benefits of Telecommunication Service Utilisation

5.1.1 Purpose of Telecommunications Utilisation

One feature of telephone calls that already provides some information about their content is their destination and origin. In this sense, Figure 5-1 reflects the importance of Accra as the most important destination and origin of telephone calls. Together with the capital, calls directed to or originating from that city account for around 80% of all calls made. Those going to destinations within the region account for another 15% with only around 6.5% coming in from this spatial category. International calls were divided up into those from other Sub-Saharan African countries and those outside (cf. Figure 5-1).

The generally high frequency of incoming calls might be due to the fact that people call the respondents or send messages. They also arrange to be called back by their relatively wealthier counterparts outside the community (AKATSI DISTRICT ASSEMBLY 1999).

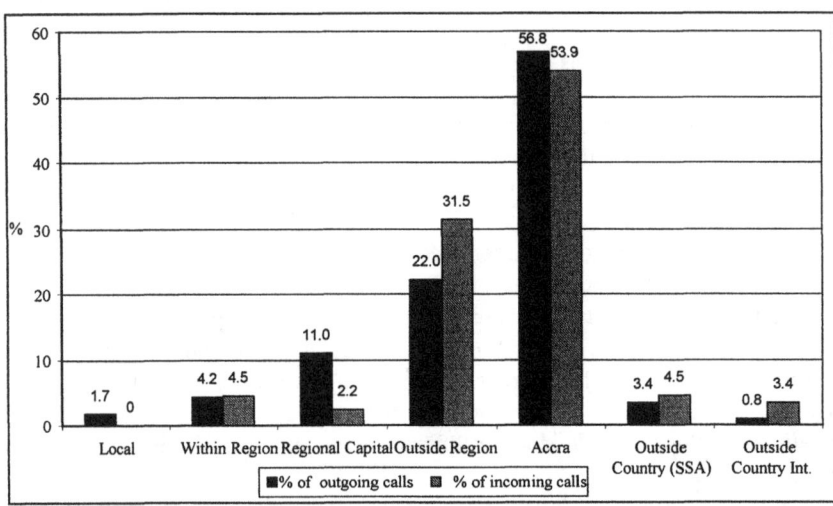

Figure 5-1: Destination and origin of outgoing and incoming calls (N = 118 for outgoing, N = 89 for incoming calls)[34]

In order to find out more about the counterparts of the individual calls as well as the purpose of the calls made and received, the interviewees were asked to explain what kind of information they were transmitting. The results generated from this question are summarised in Figure 5-2: approximately 64% of the interviewees use the services to communicate with family and friends, followed by around 16-17% that utilise phone or fax services related to the economic activities of their household. Another 7-8% of the calls were related to the transmission of emergency and health-related information. Information about governmental and administrative issues was transmitted in approx. 5% and 3.5% of all calls respectively.

---

34 One should bear in mind that a proportion of calls within the region and all calls to the regional capital are billed as local calls.

Due to the small size of the sample and the limited degree of variation in the answers, cross tabulations and other bivariate analysis cannot be carried out to show potential differences between the primary purpose of making calls and the community to which the households belong. From a less restrictive point of view, it may be worth noting that all but two calls made and received by households in Agbedrafor and Gefia are to family members or friends as well as due to an emergency or health-related issue (cf. Annex 5-1). Once again, this might reflect the generally higher penetration of non-agricultural and more market-oriented economic activities in Akatsi Town.

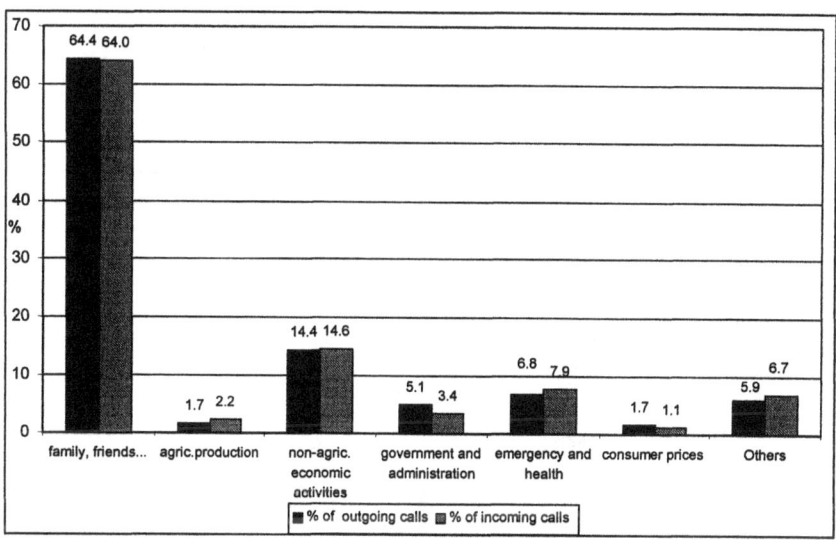

Figure 5-2: Main purposes of telecommunication use (N = 80 for outgoing, N – 61 for incoming calls)(for details cf. Annex 5-1)

5.1.2 Benefits from the Users' Perspective

As we now know more about what the respondents use telecommunication services for, the next step is to ask where the respondents themselves see specific benefits from using the phone. This was done empirically by analysing an open question that asked the interviewees to report on the three most important benefits they derive from using telecommunication services. As shown in Figure 5-3, around 33% of all answers indicate that service utilisation saved money. This is followed by the benefit of immediate bilateral conversation (28%), saving time

137

(16%) and the reduction of physical risks of travelling (10%). As could be expected from the analysis of the main purposes of the call, the immediate benefits related to business activities and market participation are relatively rare. This also holds true for the advantage of being able to increase the amount of information exchange as well as the safe and confidential way of communicating through the telephone (cf. Box 5).

Box 5: Confidentiality of telecommunication services
It was indicated by the researchers carrying out the interviews that those interviewees that value the safety and confidentiality of the tele-communication are likely to be illiterate: without access to telecommunications they would most probably ask a third person to write a letter, which is certainly less confidential than a phone call.

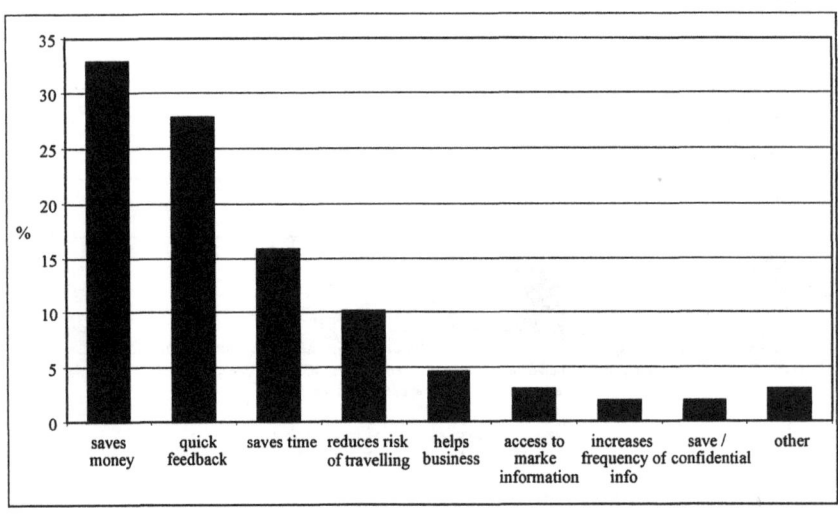

Figure 5-3: Most important benefits from telecommunication use from the respondents' point of view (up to three answers per interviewee were possible, N=205)

In line with the perception of the household heads, the key informants pointed out the benefits derived from cheaper and quicker communication. They also point out the possibility of communicating easily and directly with people in distant cities, outside Ghana, and abroad (AKATSI DISTRICT ASSEMBLY 1999; AGBEDRAFOR 1999; GEFIA 1999).

All key informants did, however, complain about the limited availability of the phone service and – especially in the case of Capital Telecom – about the high fixed costs and low quality of service: the benefits on both the household and the community level could be much higher if the services were established or were at least more accessible.

Bearing this in mind, we will now proceed to look at the ways telecommunications have an impact on how

- small-scale producers and traders might be able to enhance input supply and output marketing,
- organisations such as co-operatives, associations, and loan groups could be reached more easily by their members as well as improve their internal communication flows,
- private and public services could be more accessible and more efficient, and
- the information flows related to the administrative set-up could be improved.

These aspects will be discussed relative to the role telecommunications currently play and the role the services could potentially perform. Methodologically, the discussion is based on the key informant as well as the household questionnaires applied in the survey. The former asked the community experts about the role telecommunications play for the abovementioned groups and organisations. The latter more directly assessed the interaction of households with input, output markets and district-based commercial and public entities. It was initially planned that the data generated would be used to quantitatively compare costs of specific information flows between using and non-using households. However, the low variation in answers and the low absolute number of cases determined the qualitative nature of the analysis.

Concerning the major income-generating activities in the district, it was pointed out in previous sections that a large proportion of the population lives from agriculture and generates their monetary income from the sale of surplus food on the local markets. The spatial focus on the Akatsi market and the sheer lack of telecommunication infrastructure between Akatsi and the smaller communities in the district might explain the low importance of the medium to transact agricultural information (cf. Figure 5-2). On that note, the key informants explained that the direct economic importance of telecommunication services especially holds true for the few larger traders and market-oriented farmers: they are willing to pay for and use the services to communicate with their customers and suppliers in the urban centres. For the majority of small-scale farmers, great potential was seen in the enhancement of agricultural extension services through the use of telecommunications. If the agents of the services as well as the key farmers in the extension groups used telecommunications, a proportion of costly trips to the farmers could be avoided. Moreover, meetings could be arranged more efficiently between the farmers and the agricultural extension officers. The latter could, again, better transmit information about timing and quantity of rainfall events, the availability of inputs as well as marketing opportunities.

Beyond traders and farmers, the – mostly informal – manufacturing and service enterprises predominantly serve the local markets. This might explain the rather low proportion of telecommunication utilisation for such economic activities (cf. Figure 5-2).

The survey also tried to assess the degree to which the work of co-operatives, associations as well as saving and loan groups could be enhanced by the use of telecommunications. Only two interviewees used the phone to contact the *Ghana Private Road Transport Union* (GPRTU): one of them – the local representative of this organisation – to co-ordinate the Akatsi branch with other GPRTU offices; the other to arrange for a loan for his own transportation enterprise.

Another two respondents participated in associations and occasionally used telecommunications to communicate with them. In one case, a drugstore owner calls the chemical sellers' association to get to know about its courses that are offered to update retailers on new pharmaceutical and chemical products. In the other, a tailor gets access to information about fashion venues by contacting the relevant associations.

Neither on the household nor on the key informant level were telecommunication services linked to the financial sector, i.e. banks within or outside the district. This might indicate that the two banks in Akatsi provide a sufficient service to the public. From the perspective of the banks, however, one would hypothesise that an improved information and data exchange with the headquarters in either the regional capital or Accra might be desirable in view of the fact that any interaction to date has taken place through physical transportation.

Earlier, we identified education and health-care as the most important public goods offered. Whereas the direct interaction between the facilities providing these goods and the population in the communities may not require *telecommunication*, no indication was given by the experts as to how the interaction between the facilities themselves and official bodies on higher administrative levels is affected. If one considers the shortage of basic equipment as well as of qualified personnel in the respective facilities, preferences might be given to more pressing issues. From Figure 5-1 we do, however, know that the transmission of emergency and health-related matters is said to be one of the more important purposes of telecommunications utilisation. Due to the fact that most health facilities in the district do not have a telephone in place[35], we expect that this

---

35 The owner of a private clinic in Akatsi itself is subscribed to the Capital Telecom network.

140

relative importance is derived from the exchange of health or emergency-related news between either facilities outside the district or between family members.

Concerning the administrative bodies, it is just the district assembly that could be supplied with a telecommunication and fax connection to better collaborate with the mostly public bodies on a regional and national level. Information flows that relate to administrative issues below the district level are very much restricted to message sending and face-to-face contacts. Thus, the area councils cannot communicate with the district heads in an efficient manner (GEFIA 1999).

It was indicated earlier that the modern and traditional authorities share similar obligations in their day-to-day activities and that the different authorities should co-operate closely in order to achieve consistency. Clearly, better information flows do not necessarily reform such co-operation if they are not founded on mutual agreements and understanding. There seems to be no doubt, however, that more efficient transmission of information could enhance existing arrangements and eventually be beneficial for the administrative structure (AKATSI DISTRICT ASSEMBLY 1999).

5.1.3   Considering Family Networks

It became obvious from the previous sections that telephone utilisation is mainly concentrated on information exchange between Akatsi including its hinterland and Accra. The survey results also show that the main reason why people use the phone seems to be to communicate with family members and friends. The major benefits seen by the respondents are saving time, money, and having the possibility of quickly communicating interactively quickly. This might indicate that the transmission of information over the telephone is seen as a substitute for travelling rather than a strategic resource that improves business performance.

The latter notion might be extended to other sections of the rural society that were observed. The lack of access to telecommunications on the sub-district level does not allow commercial, public, and administrative entities to benefit from telecommunication services. This is despite the fact that telecommunications reduce the time for information to flow and leads to better and less costly decision-making and eventually to an overall enhanced organisational performance (cf. Section 2.3; LEFF 1984). In principal, this would lead to the conclusion that the utilisation of the phone is purely consumptive in its nature and that the majority of calls are made to have a chat about the well-being of a spouse, children, and other family members.

From the theory of rural organisation we do, however, know that one needs to consider the family as the basic economic unit: where formal insurance and financial markets do not exist or are not accessible, close ties among relatives and friends compensate for such shortcomings and offer, for instance, access to informal credits and social safety mechanisms (STIGLITZ 1988). Be it the matriarchal Akan or the patriarchal Ewe: for the individual members, family means economic security, help in case of emergencies, and access to production factors such as land and labour.

Whereas such aspects were – on the local level – already considered by regarding the *household* as the reference unit in this work, it seems to be necessary to regard *family* as a group of people that do not necessarily share the same compound or live in the same community, region or even the same country. The reason for this expected spatial dispersion derives from extensive migration processes driven by political, economic, or education-related reasons. This dispersion certainly affects the family relationships as a whole.

Even though some authors discuss a modification of family relationships towards a more European type of a nuclear family, strong family networks remain an important element in Ghanaian society (cf. SCHMIDT-KALLERT 1994). This means that the role of the individual and his or her familiar rights and duties are persistent even if the person leaves the community to live in the larger cities or abroad: "there are nearly no urban citizens that do not have a strong relationship to his or her home community" (SCHMIDT-KALLERT 1994; p. 171). Consequently, it is taken for granted, for instance, that a family member who lives in the capital city sends money to his family or hosts and employs even remote relatives. Reciprocally, the relatives in urban areas can expect support from their families in the rural areas with supplying storable agricultural produce (cf. SCHMIDT-KALLERT 1994). On a scale that goes beyond the borders of the country, such mutual transactions cannot necessarily be assumed. The macro and microeconomic effect of hard currency coming into the country can, however, not be underestimated: "Ghanaians living abroad contribute US$ 300 to 400 million annually by way of remittances into the Ghanaian economy and it is believed that they could do even better" (ACCRA MAIL 2001). This is an equivalent of up to US$ 20 for every single inhabitant of the country and a share of around 5% of the year 2000's GDP[36].

Box 6: Announcing funerals
Community leaders and some households pointed out the financial burden people face if a close relative dies. A significant part of the funeral costs accrue due to the fact that the widespread family members need to be informed by sending one or more persons. Widely available telecommunication services could, according to the interviewees, reduce such costs, i.e. travel expenses and time spent on this travel.

---

36   According to the World Development Report 2000 (*THE WORLD BANK 2000*).

In this sense, ICTs support the maintenance of the networks that exist between families and friends that are not only culturally and emotionally but also economically motivated.

Somewhat different from making financial and material flows more efficient, the often extensive family networks and close ties between its members generate high communication costs that could be reduced by using less expensive means of information exchange (cf. Box 6).

Whereas the above aspects make the importance of personally motivated calls in emergency situations conceptually plausible, the tool to quantitatively capture more detailed information about flows of money or goods that are expected behind *communicating with family members and friends* could not be applied satisfactorily. This is due to the fact that the respondents could not directly quantify their benefits from the information exchange with family members but – to a larger extent – were also reluctant to do so. It is, in this context, not surprising that the answers to the question about the benefits of the last telephone call made go beyond the categories produced in Figure 5-3 and were often referred to as *personal* matters. The 26 respondents that were willing to answer related questions mentioned benefits shown in Table 5-1. As apparent from the listing, most interviewees arranged financial help and remittance flows over the phone. To a lesser extent, information about business inputs and outputs were exchanged with relatives and friends. This also applies to information related to work and education opportunities and the discussion of a higher salary.

Transaction	Frequency
Financial help/remittance payments could be transferred quicker	17
Market price information could be exchanged	3
Input supply could be arranged	2
Exchange of labour force could be arranged	2
Joint business plans could be discussed	1
Quick feedback on education plans was given	1
A higher salary could be discussed successfully	1

Table 5-1: Economic transactions where telecommunications are used for family and friend-related purposes

Although by no means representative in a statistical sense, the results of this section made clear that information exchange with family members and friends living outside the communities plays an important role for the households interviewed. These communication flows cannot purely be regarded as consumptive in their nature and might have a rather significant impact on the households' welfare that goes beyond the personal satisfaction of being able to directly communicate with each other. Thus, the role of telecommunications in helping families and friends to maintain close ties may yield substantial socio-economic benefits.

5.2 Benefits from Space and Time Bridging

One conclusion drawn from Chapter 4 was that users are better off and more actively involved in economic activities outside subsistence farming. To quantify the effects of telecommunications use it might be helpful to take a closer look at the benefits seen by the users themselves. They particularly emphasise the money and time saving effect of the utilisation.

5.2.1 Conceptual Foundation

Conceptually, this leads us back to the main characteristic of ICTs: the capability to allow the decoupling of information from its physical repository over increasing geographical scales with declining unit costs (cf. Section 2.1). We concluded that this leads to the possibility of immediately transmitting information independently from physical movement. It is this property of the technologies that made scholars discuss the *death of distance* (BRUNN, LEINBACH 1991; CAIRNCROSS 1997). Whereas we do not elaborate any further on this complex and more or less philosophical question, it is important to recognise the very basic and almost trivial link between the technologies' characteristics of space and time bridging and their economic consequences.

This link can be approached using the time geography concept. According to HÄGERSTRAND (1970, 1972), any form of human activity is constrained by space and time. This limitation forces human beings to set priorities, which happens – assuming rational behaviour– in consideration of opportunity costs. The concept of limited resources is conceptually adjusted to the time and space resources available. Criticising the simplicity of highly aggregated analyses of spatial behaviour, HÄGERSTRAND developed an analytical framework that paid tribute to the individual's situation and regarded his or her life paths "as a sequence of projects (a combination of goals and tasks) represented by the creation of bundles in time-space linked by travel from one project to another" (SHANNON 1997, p. 42). The complexity towards human behaviour thus created built a milestone in human geography because it went beyond regarding individuals as consumers and labourers and considered the space and time-related limitations of individuals in caring for their everyday needs (KLINGBEIL 1980).

The assessment of the individuals' sequence of activities was generally regarded as too complex for quantitative empirical research (KLINGBEIL 1980). Nevertheless, the approach helps us to gain insights into the link between time and space bridging and the quantification of the benefits derived from telecommunications use. In the context of a rural community, this happens along the following lines:

Constraints of space and time are especially serious in rural areas of developing countries where poor transportation and other infrastructures as well as limited financial and time-related resources hinder people from overcoming the constraints. The decision to exchange information with any distant source often requires the consideration of excessive travel time and associated costs. This makes information and communication particularly expensive and limits the information flow and interaction between the various agents. This chain eventually leads to negative implications on the efficiency, productivity, and welfare of the various agents (LEFF 1984; SHANNON 1997)(cf. Figure 5-4).

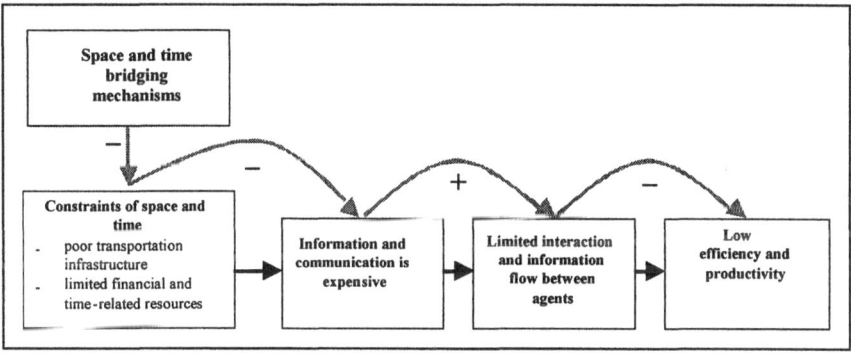

Figure 5-4: Conceptual reasoning of ICTs' time and space bridging mechanisms (cf. LEFF 1984; SHANNON 1997; HÄGERSTRAND 1972)

We can now imagine how the use of telecommunications reconfigures this chain (cf. Figure 5-4): they bring the information source or communication partner within reasonable range of the respective person or organisation. This lessens space and time constraints and leads to two, often mutually dependent, effects. On the one hand, existing information and communication channels become less expensive in terms of the time and money required to use them. On the other hand, "a movement down the demand curve for transmitting information will increase the amount of information that is sent. [...] similar effects apply to the quality (timeliness, completeness and reliability) of the information produced" (LEFF 1984, p. 258). The limiting effects on human interaction could be lifted through the complementary information which, in turn, can result in the economic benefits reported earlier (cf. Section 5.1.2; LEFF 1984).

The following section will quantify the benefits from using telecommunication services if they are applied to make existing information channels more effective (Section 5.2.2). The way in which the increase in the amount of information available enhances the households' position as a small scale producer or trader will be described in a qualitative manner in Section 5.2.3.

5.2.2 Alternative Modes of Communication and Consumer Surplus

The effect of ICTs to enable less expensive information flows leads to the question of which more expensive forms of communication can be substituted by telecommunications. SAUNDERS ET AL. (1994) specifically elaborate on those alter-

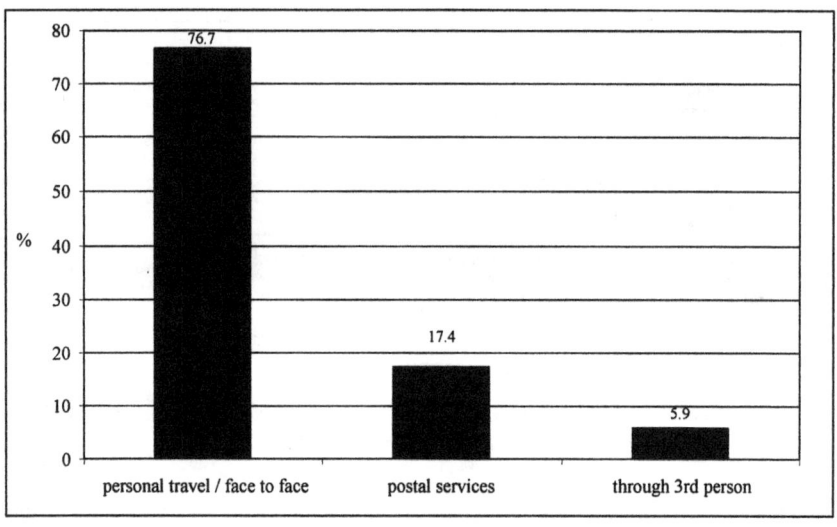

Figure 5-5: Importance of alternative means of information exchange

natives that would usually be face-to-face communication requiring physical travel, postal services as well as sending messages through third persons. Conse-

146

quently, the question asked in this section is the extent to which telecommunications can substitute these forms of information exchange.

Figure 5-5 reflects the results of asking the interviewees what mode of information exchange they would have substituted their last outgoing call for, if no telephone had been available. Around 88% of all respondents saw a possibility to substitute their last call made. Out of this group, more than three quarters claimed that they would have travelled, another 17.4% would have used some form of postal service and approx. 6% would have sent the information through another person.

To capture how much a household budget gains by using telecommunications rather than the alternative means of information exchange, we will now reconsider the motivation of our estimation exercise carried out in Section 4.3.3.

Remember, to be able to estimate whether or not the respective household members will use the service, it was crucial to calculate the difference between the utility of using the service and the cost of this utilisation. Assuming rational behaviour, the decision to eventually use the service was determined by a surplus from using the phone ($V(tc)$) as compared to the cost of that use (cf. Eq. 3). On the notion that this surplus is, at least partly, generated due to the fact that a more costly means of communication could be avoided, one could calculate the cost-benefit from telephone use. Subtracting the latter's overall costs ($C(x_{it})$) from the costs accrued through the best alternative means of communicating ($C(x_{ia})$) generates a portion of the consumer surplus ($S(x_{it})$) expressed as (SAUNDERS ET AL. 1994):

$$S(x_{it}) = C(x_{ia}) - C(x_{it}) \qquad \text{(Eq. 14)}^{[37]}.$$

The costs of using either phone services or the alternative were derived from the information given by the interviewees that participated in the household survey and are related to their last outgoing phone call. Those costs consist of the time required and the travel expenses accrued to get to either the telecommunication service or the alternative information source, staying there and going back again. Moreover, direct costs to use the telephone as well as the costs for the alternative transmission, i.e. postal services had to be included.

In order to be able to quantify the opportunity costs of time, i.e. the potential monetary loss for the user during the time the information exchange is taking

---

37   In this context it is necessary to assume the same quality of information and that both alternatives are perfect substitutes, i.e. that the user does not prefer the alternative for reasons other than costs (SAUNDERS ET AL. 1994).

place[38], the legally guaranteed daily minimum wage of ¢2.000 for unskilled work is regarded as the reference income (r), assuming an average working day contains 10 hours (cf. AKATSI DISTRICT 1999). With the data that was gathered, the cost equation leading to the partial consumer surplus:

$$S(x_{iv}) = (A(x_{ia}) + B(x_{ia}) + (D(x_{ia})*r)) - (A(x_{it}) + B(x_{it}) + (D(x_{it})*r)) \qquad \text{(Eq. 15)}.$$

Where

$A(x_{ia})$ = money spent to get to alternative (round trip)

$B(x_{ia})$ = money spent to use the alternative

$D(x_{ia})$ = time spent on round trip to alternative, including the use of the information source

$A(x_{iv})$ = money spent to get to telecommunication facility (round trip)

$B(x_{iv})$ = money spent to use the telecommunication facility

$D(x_{iv})$ = time spent on round trip to telecommunication, including the use of the facility.

Table 5-2 shows the benefits per call for those interviewees who would have substituted their last phone call for the alternatives mentioned. Whether it was the time saved, the monetary cost of communication, or the portion of the consumer surplus that was calculated, the results show high returns on using the telecommunication services as compared to the best alternative.

The comparison between those using households that are – in monetary terms – richer and those that are poorer than average shows an interesting result (Table 5-2): Slightly higher gains in absolute terms are achieved by the richer households (¢15410.4). In relative terms, however, the calculated surplus accounts for an average of 9.0% of the poorer households' monetary income and 3.2% of the richer ones respectively.

These results certainly only reflect potential savings because they neglect the fact that a phone call never can be seen as a 100% substitute for a face-to-face meeting. It would, for instance, be necessary to consider the case where the information exchange, e.g. through personal travel, is combined with other activities such as buying and selling of goods that might be cheaper in the town that was visited. Those kinds of mechanisms could not be considered due to financial and time related constraints of the survey. It is therefore not possible to monetarise the difference in personal satisfaction from a face-to-face as opposed to a *tele-conversation*. Moreover, the chosen focus on the rural households only allowed us to show the benefits that accrue for the households in and around Akatsi; it did not cover the positives from better rural-urban communication that may

---

38 Due to the fact that the respondent is not directly involved in travelling if his or her best alternative to phone use is using postal services or sending another person, the opportunity costs of time are assumed to be 0 in these cases.

be generated for the counterparts in other locations, i.e. towns and cities outside the district.

	All Using Households		Poorer Households		Richer Households	
	costs saved (in Cedis)	time saved (in min.)	costs saved (in Cedis)	time saved (in min.)	costs saved (in Cedis)	time saved (in min.)
Min	-1000	-65	-1000	-13	500	-65
Max	108000	1016	37000	837	108000	1016
Mean	13211.1	287.6	11320.3	291.2	14222.6	285.7
Std. Deviation	12212.1	186.9	7162.8	138.3	14171.8	209.2
	surplus (S) (in Cedis)	S as % of household income of last month	surplus (S) (in Cedis)	S as % of household income of last month	surplus (S) (in Cedis)	S as % of household income of last month
Min	24	n.a.	24	0	601	0
Max	109055	n.a.	42001	51	109055	14
Mean	14337.4	n.a.	12416.7*	10.1	15410.4*	3.7
Std. Deviation	12544.7	n.a.	7814.7	9.0	14510.3	3.2

*Table 5-2: Cost and time savings and consumer surplus generated through the use of telecommunication services for all households (N=68); * a t-test did not generated significant results (p>0.1)*

Nevertheless, the results presented make it seem reasonable to assume that without access to telecommunications, the demand for bilateral communication over distance would be covered most of the time by the use of more costly modes of communication. SAUNDERS ET AL. (1994, p. 148) support this finding in pointing out that in areas "where telecommunication infrastructures are highly inadequate, trips are made that could be replaced by telecommunications by any standard of judgement".

5.2.3 Information Increase, Decision-making, and Market Participation

In addition to the effects mentioned that enable a more efficient use of resources, improved and extended telecommunication systems will likely have complementary effects and increase the absolute amount of information exchanged. As already pointed out in Section 2.4.2, the relative information poverty of economic systems could be reduced through the strategic application of telecommunications. How this could look like will be analysed by first referring to conceptual thoughts discussed by LEFF (1984) and supported by anecdotes that were gener-

149

ated during the survey. This is done in a qualitative manner due to the low number of cases and the high degree of variation.

Unlike the substitutive use and in line with the conceptual reasoning summarised in Figure 5-4, the reduction of space and time constraints increases the interaction between people and the absolute amount of information available. This mechanism opens up to agents the possibility of acquiring additional intelligence about the options they have. The simple possibility of choosing from different economic alternatives already increases the basis for more rational decision-making (GEERTZ 1978; LEFF 1984). For the small-scale producers or traders, enhanced decision-making processes first and foremost positively affect their access to market information such as availability, prices, and qualities of inputs and outputs. Additionally, decreased communication costs also enable producers or traders to negotiate transactions over distance. As already pointed out (cf. Section 2.3), these kind of mechanisms are expected to induce two different phenomena, i.e. the emergence and the spread of markets: "Lower costs for acquiring information reduce the fixed-cost hurdle that must be overcome if a market for given product or input is to emerge at all. In addition better communications and lower transaction costs reduce the variable costs of market participation and operation" (LEFF 1984, p. 262).

We are, however, not considering the markets as such, but rather focusing on their (potential) participants. Related to the survey area, those agents that use telecommunications are potentially able to decide whether to stick to the local market or whether to more actively participate in more distant ones that provide a wider range of goods and better price conditions respectively. Access to telecommunications does then allow them to acquire information and negotiate transactions over distance. Eventual participation in those distant markets will depend on the market conditions found locally as compared to the markets at another place. Additionally the costs and benefits, i.e. transportation costs and opportunity costs, expected profit margins etc., associated with both possibilities need to be considered. On this note, Figure 5-6 reflects the decision-making processes that relate to telecommunications and market participation.

We will now proceed to support the somewhat abstract report of the decision-making processes with anecdotes obtained from the surveying process.

One respondent used the possibility of easy and cheap communications to decide whether to travel to a distant market or rather wait and sell the products locally. The interviewee explained that he regularly calls his brother who lives near Accra to obtain the market prices of dried peppers. As this cash crop can easily be stored and cannot be sold at profitable prices on the Akatsi market, he decides when to travel to the Ghanaian capital according to the price information obtained. Related to the input markets, a farmer in Gefia reported on the possibility of calling the supplier of fertilisers in Denu (cf. Figure 4-2) and his ability to judge whether to obtain this input from the Denu market or get it locally it from the – often more expensive – agricultural extension service.

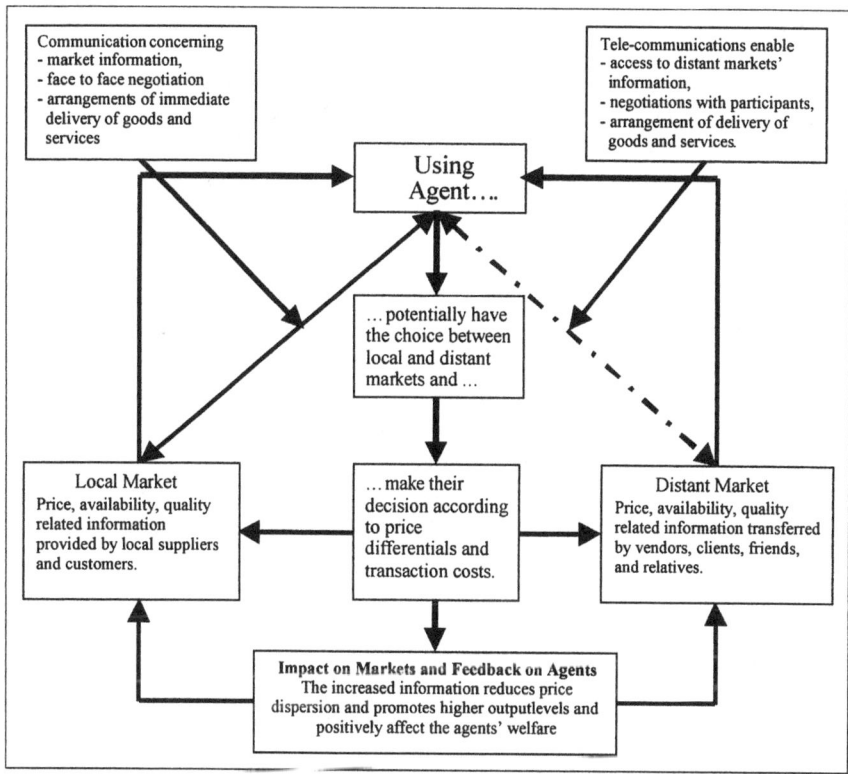

Figure 5-6: Decision making processes and market participation

Telecommunications use can also increase the position of agents who resell products that they are not able to purchase locally and facilitate the transport of the products from the distant markets. A plantain and maize trading woman, for instance, uses the telephone to obtain information about the market prices of her products at the central market in Kumasi. Approximately every two weeks, she purchases a small lorry load of plantains and 20 bags of maize, investing ¢2.9 million including transportation. On the local market the trader is able to sell the products for ¢3.1 million, earning around ¢200.000 every two weeks. According to the price of the products she negotiates on the phone, a decision is made

151

whether and when to travel to purchase the goods. Instead of travelling there personally, she usually calls a relative in the Ashanti capital to arrange purchasing and transportation. The ability to receive and make phone calls in the communication centre around the corner furthermore enables her to take measures if her goods get stuck due to a lorry break down. In those rather regular cases – due to the bad transportation infrastructure – the driver calls her or leaves a message. As a result, she is able to arrange for additional transportation preventing her goods from decomposition and eventually loosing her investment.

The effect of easy communication on the reduction of price dispersion (cf. Figure 5-6) is made transparent by another interviewee. The petty trader obtains price information from Tema and Accra and adjusts her own prices accordingly. On the one hand, this enables her to successfully compete with those products brought in from the distant markets by the commuting population or small-scale traders. On the other hand, she can increase profits if she finds out that the goods are sold at higher prices outside Akatsi.

The cases illustrated not only show that there are price and availability differentials between the Akatsi market and those markets targeted by the respondents. They also indicate that the additional information increases the market participation of the respondents and promotes their welfare. Moreover, widely spread information about distant markets also reduces the dispersion of actual prices, again leading to welfare effects for both the supplier and the consumer.

5.3 Conclusion

Besides reporting where and whom people are calling, the major emphasis of Chapter 5 was to elaborate on the benefits that are derived from using telecommunication services. From the point of view of the interviewees, those benefits are mainly cost and time savings and the possibility of interacting quickly and directly. According to the data, this interaction takes place with family and friends, which does not necessarily indicate that the purpose of the communication is purely consumptive.

Based on some conceptual thoughts around the space and time bridging characteristics of telecommunication services, we quantitatively captured potential efficiency gains by calculating a portion of the consumer surplus derived from cost differentials between telephone use and its best alternative. In doing so, we assumed that this alternative was a proper substitute for telecommunications utilisation.

The framework provided by LEFF (1984) also led us to recognise information as a strategic input into the interviewees' economic activities that enables choices and beneficial decision-making processes. We could only provide anecdotal evidence for such a complementary utilisation of telecommunication services. The examples stated made clear, however, what potential is inherent in the ability to communicate over distance at low costs – even at the level of small-scale producers and traders.

Bearing in mind those benefits, one may ask why there is not a wider array of people using telecommunication services in a more business-related and strategic manner. One reason might be that there is a lack of knowledge about opportunities that are inherent in the mechanisms mentioned. Another explanation lies in the fact that subsistence farmers and government employees make up a significant proportion of the communities' inhabitants. Both groups are not necessarily market-oriented and might stick to the local markets to sell the surplus produce. The local markets are, furthermore, dominated by the periodic Akatsi market that is taking place every five days and whose importance goes beyond the borders of the Akatsi district. The completeness of market information for many products created in this manner is also likely to reduce the need for information from outside the district capital.

Another approach towards explaining the reluctance to acquire new information from outside the locality is associated with the increased probability of receiving false information. Apart from the notion that "there seems to be no a priori basis for presuming that modern telecommunications reduce the ration of true to false messages" (LEFF 1984, p. 270) the survey results do not support these fears. Both the expert interviews as well as the qualitative information captured do not indicate such negatives.

Concerning the value and reliability of information coming in, one can also imagine a scenario – as with the plantain trader or the pepper farmer – in which new market level information is acquired from relatives and friends. The issue of the role of social networks and how they guarantee the trustworthiness of information does raise more questions in this respect, which point out the need for information-flow related research in rural areas of low income countries.

6 Concluding Remarks

The genuine excitement about digital technologies and their potential to promote economic development seems to leave at least two important aspects uncovered. The first is that the availability of the technologies per se does not create welfare. The second points out the discrepancy between the rapid diffusion of the latest technologies in urban areas of many developing countries and the lack of even the most basic telecommunication infrastructure in their rural areas. In this light, the Maitland Report, entitled *The Missing Link*, from the mid 80s, which urged for bringing the people within easy reach of a telephone, gets a new dimension (G8 2001; MANSELL 1999; INDEPENDENT COMMISSION FOR WORLD-WIDE TELECOMMUNICATION DEVELOPMENT 1984). It was one important task of this work to point at this dimension by showing political and regulatory trends that foster ICT implementation on the national and sub-national level. On the latter, we addressed the scarcity of empirical work as soon as it comes to rural communities and households and tried to assess who is currently using telecommunication services in rural areas, what this usage looks like, and how the benefits from the use can be characterised. This task was followed by conducting a survey in Southern Ghana. Although the limited range of the survey does not allow generalisation in a strict statistical sense, the universality of the methods applied and questions asked allows the evaluation of the result in a more general manner, eventually leading us to policy and research-related implications.

6.1 Summary of the Results

A foundation to answer the major research questions was provided in Chapter 2, which gave an overview about the major determinants and dimensions of ICT diffusion as well as a review of how ICTs and economic development interrelate. The wide array of concepts underlined the need for a more detailed empirical assessment on the grass-root level; it also showed the necessity to narrow down the scope of the study to publicly accessible telecommunication facilities.

The question of who actually uses the services was then addressed by axiomatically considering huge shortcomings in the supply of telecommunications infrastructure in low-income countries and, hence, the ability to neglect the existence of proper market mechanisms. In this context, we elaborated on the current

155

trends in many low-income countries to address this failure and build-up an information and communication infrastructure that is beneficial for a larger proportion of the population. The tools to do so are already implemented in numerous low-income countries and in line with those applied to industrialised economies. The tools mainly comprise privatisation, liberalisation and regulatory strategies.

How Ghana's internationally acknowledged sector reform is manifesting itself on the institutional and the infrastructural level was described and shortcomings were discussed. We recognised the persistent market power of the former monopolist and its limited willingness to fulfil the licensing agreements. The regulatory body was reinforcing the problems by being too passive and not endowed with sufficient capacity to overcome its weak position.

The assessment of the infrastructural developments that came along with sectoral reform showed that, in order to maintain its market power, Ghana Telecom significantly expanded its network in the cities. It seems, though, as if the company cannot cope with its commitment to successively equip every community above 250 inhabitants with at least one public telephone line. One reason for this lies in the focus of the company on first of all expanding its network to areas of higher population density and economic activity. On that basis, it became obvious that, despite the sector reform, most rural areas and even many district capitals do not yet have a telecommunication infrastructure in place. This physically excludes a large proportion of the population from using telecommunication services.

This situation was confronted with an assessment of access and use on a different scale. Taking three communities in the Akatsi District as an example, determinants and degree of usage on the level of households in a scenario were public telephone facilities were available, were identified. The results of the empirical assessment pointed out the considerable catchment area of the services and the high penetration of use within the communities that had the services in place or were close-by. Additionally, it became clear that people are willing to take into account considerable costs in terms of time and money in order to be able to utilise the communication services. The analysis of the determinants of service use on the household and the personal level indicated that there is a high hidden demand and that this demand is by no means limited to the economically wealthy and active. However, user rates were particularly low amongst women, illiterates, and those living in significant distance to communication service facilities.

The analyses of the purpose and benefits that can be expected if telecommunications are used showed that the functioning of the local administration, as well as the provision of public services are not yet affected.

On the household level, we derived a somewhat ambiguous picture. At first glance, the time geography approach, in line with the users' perception, allowed a quantification of private time and money-related savings deriving from telephone use. At second sight, the information transmitted over the services was mainly related to personal matters. Considering the strong social ties between family members and friends as well as their social and economic relevance allowed a qualitative interpretation of what kind of economy-wide effects may hide behind *calling family and friends*. It is a clear shortcoming of the assessment tools developed in this work that they did not capture such *hidden* benefits in a more comprehensive manner, even if many interviewees were reluctant to give insights into the messages they transmitted anyway.

An increase in available information and its impact on decision making and marketing processes could not be analysed in a structured manner, either. Anecdotal evidence, however, illustrates that more and better information is an important input into the households' economic activities: an increase in information available clearly opens up choices and beneficial decision-making processes.

6.2 Evaluation of the Results – Policy Implications and Further Research Questions

Confronting the economic expectations as well as the political and infrastructural developments with the empirical part of this work indicates the huge discrepancy between political goals and infrastructural and economic facts on the ground.

The results from the survey location show that political commitment, institutional change, and technical developments have not yet trickled down to the level of rural communities and their population. This is despite the creation of an ICT strategy targeting at very ambitious goals, including universal access. One might, against this background, be disappointed and ask for the sense of sector reform if the poorer sections of the population cannot yet properly benefit from it.

This disappointment might, on the one hand, be explained by an impatient attitude resulting from the simple fact that many perceive ICT sector developments as a revolution. It should be obvious that time is required for sectoral reform to facilitate benefits for the users. On a technology and human capacity-related basis, it should also be accepted that it is a rather complex infrastructural task to connect rural areas to national networks. This also holds true if potent foreign investors enter using the latest wireless technologies.

On the other hand, the adoption of the technologies may – at the level of rural households – not yet have reached a stage that fulfils the expectations raised by policy makers. It seems as if the new technologies primarily enhance traditional means of information exchange by a more cost-effective mode of communication rather than create more economic activity through, e.g., strategic input and output marketing. Despite the fact that substitutional effects may yield substantial direct benefits, a strategic application of the services would usually emerge through an evolutionary (learning and market creation) process (cf. Sections 2.3, 5.2.3).

Both time-related issues point out limitations of this work that exclusively draws on cross-sectional data: aspects that point at time lags in the diffusion of infrastructural means and the adoption of the ICTs as strategic input factors can only be covered over time and could, thus, not be taken into consideration.

Apart from the evolutionary effect of time, clear needs for improvements in sector policy can be identified and lead to some policy-related conclusions. They particularly relate to the Ghanaian situation but might easily be transferred to the many low-income countries that have embarked on similar sector strategies.

From the perspective of infrastructural development, financial means and regulatory improvements should particularly be used to overcome the investment vacuum in rural and remote areas. Instruments should be created that promote the access of non-served communities and creative forms of access.

On the district and the community level, telecommunication co-operatives – as they were successfully introduced in many Latin American countries – could form a basis to get the infrastructure in place. The interface to the end-users might be realised through the promotion of telecentres that operate on a franchising basis and may motivate small-scale entrepreneurs to resell the services (cf. Box 3). The franchising conditions should, in this respect, be adjusted to the demand and turnover that may be significantly different in a district capital as opposed to a farming-based community of just around 400 inhabitants. The adjusted national information and communication plan, due to be launched by the new government in August 2001, seems to contain such strategies (GHANWEB 2001a).

In case cellular telecommunication providers should leapfrog the expansion of the fixed line network of Ghana Telecom, a scheme similar to the village pay phone project in Bangladesh (cf. BAYES, VON BRAUN 1999) could be introduced. Private individuals would then be able to lease or buy cellular telephones and resell the service to the public in their communities.

Network expansion should also focus on connecting public administration, health and education facilities. Information exchange between those entities, but also between the population and the public sector, may yield huge external effects. The Ghanaian government and the donor organisations involved should, for instance, be aware that their decentralisation efforts could significantly be strengthened by enhanced information flows across the various administrative levels.

The end-users' ability to utilise the telephone for marketing reasons does primarily depend on his or her ability to participate in markets but also his or her awareness of the presence of more and better information available beyond the traditional sources. Raising such an awareness does require time, but could also be promoted by the involvement of intermediaries such as agricultural extension officers.

Intermediaries are even more important when it comes to the use of ICT applications that involve computers, such as email or the WWW. They can bridge the gap between the technology and the end user. Successful applications of such mechanisms do already exist in form of the Information Village Project that was implemented in Pondicherry, India by the M.S. Swaminathan Research Foundation (cf. THE HINDU 1999). A basic precondition for those models is, however, that information relevant for people in the respective community is gathered and provided.

Research needs are obvious on two levels:

The first addresses the need to consider time as a crucial factor when it comes to the impact of ICTs on economic development. This especially holds true if one considers the rapid changes in the sector. Furthermore, households and families were considered as a black box. Future research should approach the telecommunication use and benefits with a stronger focus on the individual user and also consider the technologies' role in strengthening the important family networks. This does make it necessary to centre research more around the information content that is transferred rather than on the *technical* side of telephone use.

On a different level, governance issues should be tackled in more detail and point at deficiencies in the regulatory and institutional set-up of many developing countries' ICT sectors. Those are often characterised by a too simplistic adaptation of strategies that were originally developed in and for the industrialised world.

In essence, there is a need to call for a more adjusted and differentiated view of the potentials which are coming along with ICTs' implementation in low-income countries and the risk of excluding vast majorities from these potentials. This should not only enable a more fruitful discussion with critics that perceive the issue – in light of the often overwhelming problems with hunger, water scarcity, and physical threat – as cynical. It should, moreover, foster sustainable and on-the-ground developments and applications that consider the importance of basic telephony. This tool, which to many seems less fascinating when compared to the internet, still remains *the* missing link for the vast majority of the people in most low-income countries.

Abstract

Focusing on Sub-Saharan Africa and Ghana in particular, the overall goal of this research is to empirically assess how Icts are accessed and used by rural households and how the latter benefit from this use. To achieve this goal it is, on the one hand, necessary to analyse the economic and political framework that determines the access to information and communication technologies (Icts). On the other hand, the socioeconomic characteristics of the households that lead to the use of the infrastructure are crucial elements of this work.

Besides providing an overview about the existing literature that discusses the link between ICT diffusion and use and measures of economic development and welfare, the general trends of ICT sector reform as observable in many low-income countries are discussed.

In the Ghanaian context, the persistent market power of a former monopolist, its efforts to expand the profitable urban ICT infrastructure and the subsequent exclusion of rural areas can be identified as a problem: despite a significant sector reform, even many district capitals do not yet have a telecommunication infrastructure in place. This has restricted the assessment of the current use of Icts to the most basic ones, namely publicly available telecommunication services.

To show how households use – and benefit from – telecommunication services, determinants and degree of usage in a scenario where public telephone facilities are available have been analysed. The results of the empirical assessment in three villages in Southern Ghana show the considerable catchment area of the services and the high penetration of use within the communities that had the services in place or were close-by. Additionally, the research uncovers that people are willing to take high costs in terms of time and money into account in order to be able to utilise the communication services. The analysis of the determinants of service use on the household and the personal level indicates that there is high hidden demand and that this demand is by no means limited to the economically wealthy and active. However, user rates are particularly low amongst women, illiterates, and people living in significant distance to public service facilities.

The examination of the purpose of use and the benefits that can be expected from it show that the functioning of the local administration, as well as the provision of public services in the survey area, are not yet affected by telecommunications. On the household level, a somewhat ambiguous picture can be drawn. At the first sight, the findings allow a quantification of private time and money-related savings deriving from telephone use. At second, the information transmitted over the services appears to be mainly related to personal matters. Considering the strong ties between family members and friends as well as their social and economic relevance for an individual allows a qualitative interpretation of what kind of economy-wide effects may hide behind *calling family and friends*.

Anecdotal evidence from households and family businesses let us further recognise more and better information as important inputs into the households' economic activities: an increase in information available clearly enables choices and beneficial decision-making processes.

Confronting the economic expectations and political vow with the empirical part of this work indicates the huge discrepancy between political goals and infrastructural and economic facts on the ground: the results from the survey location show that political commitment, institutional change, and technical developments have not yet trickled down to the level of rural communities and their population.

Against this background the following needs can be identified:

From the perspective of infrastructural developments, financial means and regulatory improvements should particularly be used to overcome the investment vacuum in rural and remote areas. Instruments should be created that promote the access of unserved communities to national telecommunication networks. Network expansion should also focus on connecting public administration and services. Information exchange between public entities, but also between the population and the public sector, may yield huge external effects.

On the level of the end-user, his or her ability to utilise the telephone for marketing reasons does primarily depend on market participation, but also on his or her awareness of the availability of more and better information. Raising such an awareness does require time, but can be promoted by the involvement of intermediate sources such as the agricultural extension service, co-operatives and associations.

In essence, there is a need to call for a more adjusted and differentiated view of the potentials which are coming along with ICTs' implementation in low-income countries and the risk of excluding vast majorities from these potentials. This should, moreover, foster sustainable and on-the-ground developments and applications that consider the importance of basic telephony as a tool to foster economic development.

Annex

Annex 3-1: Telecommunication regulation situation (Source ITU 1998, BMI-T 1998)

	Separation of PTO[1]	Independent Regulator[2]	Operator Privatised[3]	Int. Partners of National Carrier	Cellular Competition[4]
Angola	Angola Telecom (1985)				
Benin					
Botswana	BTC (1980)	BTA (1996)			1998
Burkina Faso	Onatel (1994)				
Burundi	Onatel (1979)	ARCT (1997)			
Cameroon					
Cape Verde	Cabo Verde Telecom 1995	DGC (1992)	1996		
Central Afr. Rep.	Socatel (1982)		1990: 40%**	FranceCables et Radio (FCR)	
Chad			1976:48%*	FCR 43%, Alcatel Telspace 5%	
Comoros					
Congo					
Côte d'Ivoire	CI-Telecom (1995)	AT-CI (1996)	1997: 51%	FCR	1996
DR Congo					
Djibouti			1977:25%*	n.a.	
Equatorial Guinea			40%*	n.a.	
Eritrea	TSE (1993)	CD (1996)			
Ethiopia	ETC (1967)	ETA (1996)			
Gabon			39%*	n.a.	

1   Indicates whether postal and telecommunications activities have been separated. If so, the year separated and the name of the principal telecommunication operator is shown. Note that the name may have changed since posts and telecommunications were separated.

2   Indicates whether an independent telecommunication regulatory agency has been established. If so, the name and year the agency was created is shown.

3   Indicates whether the telecommunication operator is private. If so, the date and per cent sold is shown. An asterisk (*) refers to the international operator. Two asterisks (**) refer to "privatisation" through the fusion of the partly-private international operator and the state-owned national operator.

4   Indicates whether there is more than one network operator providing cellular service.

Gambia	Gamtel (1984)				
Ghana	Ghana Telecom (1995)	NCA (1997)	1996: 30%	Telecom Malaysia Bd.	1995
Guinea	Sotelgui (1992)	DNPT (1996)	1995: 60%	Telecom Malaysia Bd.	1995
Guinea-Bissau	Guiné-Telecom		1989:51%	n.a.	
Kenya					
Lesotho	LTC (1980)			ds	
Liberia					
Madagascar	Telma (1994)	OMERT (1997)	1995: 34%**	n.a.	1998
Malawi					
Mali	Sotelma (1989)				
Mauritania					
Mauritius	Mauritius Telecom (1988)	TA (1988)			1996
Mozambique	TDM (1981)	INCM (1992)			
Namibia	Telecom Namibia (1992)	NCC 1995			
Niger	Sonitel				
Nigeria	Nitel (1985)	NCC (1993)			
Rwanda	Rwandatel (1992)				
S.Tomé	CST (1982)		1989: 51%		
Senegal	Sonatel (1985)		1997:61%	FCR	
Seychelles	Cable & Wireless		100%	n.a.	
Sierra Leone	SLNTC (1995)				
Somalia					
South Africa	Telkom (1991)	SATRA (1997)	1997:30%	SBC Communications Int., Telecom Malaysia Bd.	1994
Sudan	Sudatel (1994)	NCT (1996)			
Swaziland					
Tanzania	Tanzania Telecom (1992)	TCC (1994)			1996
Togo	Togo Telecom (1996)				
Uganda		UCC (1997)			
Zambia	Zamtel (1994)	CA (1994)			1997
Zimbabwe	PTC (1998)				1998

Annex 3-2: Figures of telecommunication infrastructure within the regions
(Source: GHANA TELECOM 1998)

Region	Regional capital	Approx. number of main lines in regional capital	Proportion of regional cap. lines of all lines in region (%)	Connected main lines within region	Teledensity within region
Greater Accra	Accra	115300	92.8	124230	4.27
Upper East	Bolgatanga	560	62.2	900	0.10
Central	Cape Coast	2500	67.8	3690	0.23
Volta	Ho	710	48.8	1455	0.09
Eastern	Koforidua	2300	47.9	4797	0.23
Ashanti	Kumasi	15000	95.0	15790	0.50
Western	Sekondi	5800	88.5	6555	0.36
Brong Ahafo	Sunyani	850	49.4	1720	0.09
Northern	Tamale	1450	86.3	1681	0.09
Upper West	Wa	830	100.0	830	0.14

Annex 4-1: Charges and tariffs of Ghana Telecom and Capital Telecom
(Source: GHANA TELECOM 1998; CAPITAL TELECOM 1999)

Ghana Telecom	Fixed Costs	Call Tariffs (in Cedis) 6a.m. to 6p.m / 6p.m.		6a.m.	
Connection charge	¢300.000	Local*:	200 /	4 min../	5min.
		National 1*: <32 km	200 /	2,5min./	4min..
Monthly line rental	¢2.500	National 2*: 32-80 km	200 /	1,5min./	2min.
		National 3*: >80 km	200 /	1min./	1,5min.
		Cellular*	800 /	1min.	
Deposit for IDD	¢300.000	International 1:		2.600 / 1min.	
		International 2:		3.000 / 1min.	
		International 3:		2.000 / 1min.	

165

Capital Telecom	Fixed Costs	Call Tariffs	7a.m.-to 4p.m.	to 10p.m.	to 7a.m.
		(in ¢ / min.)			
Connection charge	approx. ¢900.000.	Within hub:	500	400	120
Monthly line rental	¢10.000 / month	National:	600	500	400
Monthly equipment rental	¢20.000 / month	Cellular:	900	900	900
		International 1:	2.600		
Deposit		International 2:	3.600		
National	¢30.000	International 3:	3.200		
		International 4:	4.000		
IDD	¢300.000	International 5:	4.600		

Annex 4-2: Tabular data analysis of communication centre expert interviews (Data source: TAP 1999 and telecommunication centre expert interviews)

Table 1: Source of financing the enterprise

Name of the enterprise	Source 1	Source 2	Source 3
Akatsi Premier Communications Centre	own funds/resources	bank	relatives
Zutako Communications Centre	,	,	,
Zutako Communications & Business	,	,	,
West Falia Communications Centre	,	,	,
Benak Trading and Communications Centre	own funds/resources	,	,

Table 2: Reason for establishment at this location

Name of the enterprise	Reason mentioned
Akatsi Premier Communications Centre	next doors to the owner's house, establish it after retirement and to provide telecom service even at night
Zutako Communications Centre	n.a.
Zutako Communications & Business	n.a.
West Falia Communications Centre	it is in the central Business district of the town
Benak Trading and Communications Centre	sited there because of he Aflao road

Table 3: Ownership and Management Situation

Name of the Enterprise	Managed by	Sex of owner	Sex of manager	Number of em-ployees[*]
Akatsi Premier Communications Centre	Owner	male	female	2
Zutako Communications Centre	Employee	male	male	2
Zutako Communications & Business	Employee	male	male	2
West Falia Communications Centre	Employee	male	male	1
Benak Trading and Communications Centre	Employee	male	male	1

Table 4: Number of main lines and range of telecommunication services offered

Name of enterprise	Infrastructure	Range of telecom services		
	Number of main lines	Outgo-ing calls	Incom-ing calls	Fax service
Akatsi Premier Communications Centre	2	yes	yes	yes
Zutako Communications Centre	1	yes	yes	no
Zutako Communications & Business Centre	1	yes	yes	no
West Falia Communications Centre	1	yes	yes	no
Benak Trading and Communications Centre	1	yes	yes	no

Table 5 Typical opening hours of the facility

Name of enterprise	Opening hours
Akatsi Premier Communications Centre	6.00am-6.00pm
Zutako Communications Centre	7.00am-9.30pm
Zutako Communications & Business	7.00am-9.00pm
West Falia Communications Centre	6.30am-9.00pm
Benak Trading and Communications Centre	7.00am-9.00pm

---

[*]   including manager

167

Table 6: Attraction of customers, satisfaction with GT's services and expansion plans

Name of enterprise	Attraction of customers	Satisfied with GT's service?	Plans to expand the business?
Akatsi Premier Communications Centre	quality service e.g. comfortable seats, TV, courteous reception, newspapers, etc.	yes	yes
Zutako Communications Centre	courteous reception	yes	yes
Zutako Communications & Business	courteous reception	yes	yes
West Falia Communications Centre	courteous reception	yes	yes
Benak Trading and Communications Centre	courteous reception	yes	yes

Table 7: Major success stories and problems mentioned

Name of enterprise	Major success	Problems 1	Problem 2
Akatsi Premier Communications Centre		no cooperation among owners of communication centres	having to travel to Aflao to pay bills every month
Zutako Communications Centre	helping people to organise their business activities easily	customers make calls that they cant pay for or feel sometimes that the charges are too high	
Zutako Communications & Business		have problems with customers when they receive their incoming calls	customers fighting over charges, asking for reduction etc
West Falia Communications Centre		customers fighting over charges, asking for reduction etc	
Benak Trading and Communications Centre		no problems	

168

Annex 4-3: Calculation of weighting variable (Source: BÜHL, ZÖFEL 1996)

$$W = \frac{R_{hs}}{R_{sc}},$$

where R_{hs} is the ratio of users or non users of the household survey and R_{sc} of the screening survey respectively.

Annex 4-4: Frequency of service utilisation

	Outgoing	Incoming
Average frequency of use per year	136.2	134.3
Average frequency of use per month	11.3	11.2
Average frequency of use per week	2.6	2.6

Annex 4-5: The equation used by Mitchell and Torero to maximise the utility in Eq. 1 (MITCHELL 1978, TORERO 2000) can be expressed as:

$$y = x + \partial(L + pq),$$

where L is the monthly subscription fee for telephone service, and p is the vector of prices for the various call rates. $\partial = 1$ if the consumer subscribes to telephone services and $\partial = 0$ if he or she does not.

Annex 4-6: The probit model (cf. GUJARATI 1995; MADDALA 1983)

The Probit analysis model assumes that there an underlying response variable $y_i{}^*$ exists that is defined by the following regression relationship:

$y_i{}^* = \beta'x_i + u_i$ where u_i is $\sim IN(0,1)$
(Eq. A.1)

Where $y_i{}^*$ is an unobservable or latent variable. The observed variable is a dummy variable y_i defined as

$y_i = 1$ if $y_i^* > 0$, the household uses telecommunications, and

$y_i = 0$ otherwise.   (Eq. A.2)

From the relations in Eq. A.1 and A.2 we are able to obtain

$$\text{Prob } (y_i = 1) = \text{Prob } (u_i > -\beta'x_i)$$
$$= 1 - F(-\beta'x_i) \qquad \text{(Eq. A.3)}$$

$$\text{Prob } (y_i = 0) = \text{Prob } (u_i \leq -\beta'x_i)$$
$$= F(-\beta'x_i) \qquad \text{(Eq. A.4)}$$

where F is the cumulative distribution function for u.

The probit maximum likelihood function that is used for the estimation is then:

$$L = \prod_{yi=0} F(-\beta' x_i) \prod_{yi=1} [1 - F(\beta' x_i)] \qquad \text{(Eq. A.5)}$$

F's functional form in Eq. A.5 bases on the assumption that u_i in Eq A.5 is normally distributed. If this is the case, this is the cumulative density of u_i from $-\beta'x_i$ to infinity, which is equal to the cumulative density function from minus infinity to $+\beta'x_i$. The probit model assumes that u_i is $IN(0, \sigma^2)$ and σ is normalised to be equal to 1. In that case it is possible to obtain the following form:

$$F(-\beta'x_i) = \Phi(-\beta'x_i) = \int_{-\infty}^{-\beta'xi/\sigma} \frac{1}{(2\pi)^{1/2}} \exp\left(-\frac{t^2}{2}\right) dt \qquad \text{(Eq. A.6)}$$

Annex 4-7: Goodness of fit measures and detailed results of the probit estimation

Alternative 1
Probit estimates Number of obs = 167
Wald chi2(8) = 73.23   Prob > chi2 = 0.0000
Log likelihood = -67.076799   Pseudo R2 = 0.3304

		Observed			
Predicted		0	1	Total	Percentage
0	N	26	11	37	70.3%
1	N	24	109	133	82.0%
Total	N	50	120	170	78.2%

Alternative 2
Probit estimates Number of obs = 167
Wald chi2(9) = 55.86   Prob > chi2 = 0.0000
Log likelihood = -69.576013   Pseudo R2 = 0.3054

		Observed		Total	
Predicted		0	1	N	Percentage
0	N	27	11	38	71.1%
1	N	23	109	132	82.6%
Total	N	50	120	170	77.7%

Alternative 1

	Probit Coefficients		Marginal Effects		Sig.	
	Coef.	Robust Std. Err	dF/dx	Robust Std. Err	z	P>z
sex	-.7514301	.2621971	-.2424003	.0873132	-2.866	0.004
age	-.0168066	.0088672	-.005116	.0026775	-1.895	0.058
edu_prim	.7094961	.3560523	.2050924	.0976226	1.993	0.046
edu_seco	.5648819	.4544441	.1473076	.1003003	1.243	0.214
edu_post	.6759535	.4164619	.1717017	.0894068	1.623	0.105
lnhhinc	.68562	.2088446	.2087056	.0631221	3.283	0.001
dumagbe	.2102254	.3544269	.0612721	.0980511	0.593	0.553
dumgef	-.4677471	.334229	-.1552482	.1192659	-1.399	0.162
_cons	-7.22219	2.666541	n.a.	n.a.	-2.708	0.007

Alternative 2

	Probit Coefficients		Marginal Effects		Sig.	
	Coef.	Robust Std. Err	dF/dx	Robust Std. Err	z	P>z
sex	-.7905281	.2567606	-.261563	.086028	-3.079	0.002
age	-.0127317	.0092188	-.0039914	.0028877	-1.381	0.167
edu_prim	.7260476	.3437366	.2160738	.0962428	2.112	0.035
edu_seco	.7607155	.4252081	.1940599	.0855625	1.789	0.074
edu_post	.8025569	.421351	.2039191	.0863314	1.905	0.057
richer_a	.3884249	.56953	.1074597	.1345818	0.682	0.495
poorer_a	-.6071938	.2845948	-.2063459	.101627	-2.134	0.033
dumagbe	-.03576	.3031677	-.0112843	.0962819	-0.118	0.906
dumgef	-.6748846	.3308015	-.2349069	.1236071	-2.040	0.041
_cons	1.235782	.6560508	n.a.	n.a.	1.884	0.060

Annex 4-8: Synoptic view of the variables used in the different estimations and their hypothesized impact. Where signs are doubled a stronger relationship is expected.

Variable Name	Description	Alternative 1		Alternative 2	
Regressants: characteristics of telecommunications utilisation					
USE	Telecommunication service use	increase of propensity of service utilisation			
		Expected Signs	Measured Signs	Expected Signs	Measured Signs
Regressors: individual characteristics					
SEX	Sex of household head	–	–***	–	–***
AGE	Age of household head	–	–*	–	–
EDU_PRIM	Level of education achieved	+	+**	+	+**
EDU_SECO	by household head	+	+	+	+*
EDU_POST		+	+	+	+*
Regressors: household and community characteristics					
LNHHINC	Household's expenditure during the last month	+	+***		
RICHER_A	Economic status within			+	+
POORER_A	community (self-assessment)			–	–**
DUMAGBE	Community where house-	–	+	–	–
DUMGEF	hold is situated	–	–	–	–**

* significant at p < 0,1   ** significant at p = 0,05   *** Difference significant at p = 0,01

Variable Name	Description	Alternative 1		Alternative 2	
Regressants: characteristics of telecommunications utilisation					
LNTCEXP	Intensity of telecommunication services utilisation, measured as expenditure on the services during the month before the interview	increase of intensity		increase of intensity	
		Expected Signs	Measured Signs	Expected Signs	Measured Signs
Regressors: individual characteristics					
SEX	Sex of household head	–	–***	–	–***
AGE	Age of household head	–	–**	–	–
EDU_PRIM	Level of education achieved	+	+***	+	+***
EDU_SECO	by household head	+	+*	+	+**
EDU_POST		+	+*	+	+**
Regressors: household and community characteristics					
LNHHINC	Household's expenditure during the last month	+	+***		
RICHER_A	Economic status within			–	+
POORER_A	community (self-assessment)			+	–**
DUMAGBE	Community where the	–	+	–	–
DUMGEF	household is situated	–	–**	–	–***
Tobit specific regressors					
INCS_AGR	Main source of the house-	–	–	–	–
INCS_GOV	hold's income	+	+	+	+
INCS_OTH		–	–	–	–
NOHHMEM	Number of household members	+	+	+	+*
PART_HH	Participation in local organisations	+	+	+	+

* significant at p < 0,1   ** significant at p = 0,05   *** Difference significant at p = 0,01

Annex 4-9: The tobit model (GUJARATI 1995; Maddala 1983)

The basic form of the tobit model can be shown as:

$$y_i = \beta' x_i + u_i \quad \text{if} \quad \beta' x_i + u_i > 0 \qquad \text{(Eq. A.7)}$$
$$y_i = 0 \text{ otherwise}$$

With β being a vector $(kx1)$ of unknown parameters and x_i reflecting a vector $(kx1)$ of constants. As already stated, u_i are residuals that are assumed to be independent and normally distributed, with mean zero and a common variance (MADDALA 1983).

For those observations (y_i) that are zero, thus for the non-using households:

$$\text{Prob}(y_i = 0) = Prob(\,u_i < -\beta' x_i\,) = (1 - F_i) \qquad \text{(Eq. A.8)}.$$

For all observations y_i that are greater than zero, thus for the non-using households:

$$\text{Prob}(y_i > 0) * f(y_i \mid y_i > 0) \;=\; F_i \; \frac{f(y_i - \beta' x_i, \sigma^2)}{F_i}$$

$$=\; \frac{1}{(2\Pi\sigma^2)^{1/2}} e^{-(1/2\sigma^2)(y_i - \beta' x_i)^2} \qquad \text{(Eq. A.9)}.$$

The corresponding likelihood function, can then, according to MADDALA (1983) expressed as:

$$L = \prod_0 (1 - F_i) \prod_1 \frac{1}{(2\Pi\sigma^2)^{1/2}} e^{-(1/2\sigma^2)(y_i - \beta' x_i)^2} \qquad \text{(Eq. A.10)}.$$

In the following equation this unfolds, for the former product N_0 being observations for those $y_i = 0$ and the latter being over N_1 observations for those $y_i > 0$.

$$\text{Log } L = \sum_0 \log(1 - F_i) + \sum_1 \log\left(\frac{1}{(2\Pi\sigma^2)^{1/2}} \right) - \sum_1 \frac{1}{2\sigma^2}(y_i - \beta' x_i)^2$$

$$\text{(Eq. A.10)}.$$

Annex 4-10: Detailed results of the tobit estimation

Alt. 1	Tobit Coefficients		Marginal Effects[5]		Sig.	
	Coef.	Std. Err	dF/dX	Std. Err	t	P>t
sex	-2.861864	.8887017	-2.861864	.8887017	-3.220	0.002
age	-.058514	.0285072	-.058514	.0285072	-2.053	0.042
edu_prim	3.579792	1.262319	3.579792	1.262319	2.836	0.005
edu_seco	2.784674	1.54874	2.784674	1.54874	1.798	0.074
edu_post	2.993444	1.647255	2.993444	1.647255	1.817	0.071
incs_agr	-.4660754	.9794701	-.4660754	.9794701	-0.476	0.635
incs_gov	1.193293	1.262673	1.193293	1.262673	0.945	0.346
incs_oth	-1.807188	1.741143	-1.807188	1.741143	-1.038	0.301
lnhhinc	2.277078	.6237518	2.277078	.6237518	3.651	0.000
nohhmem	.1885548	.1245833	.1885548	.1245833	1.513	0.132
part_hh	.3917128	.5348763	.3917128	.5348763	0.732	0.465
dumagbe	.1597895	1.024162	.1597895	1.024162	0.156	0.876
dumgef	-2.378048	1.174898	-2.378048	1.174898	-2.024	0.045
_cons	-22.52781	7.984212	-22.52781	7.984212	-2.822	0.005
Goodness of Fit	Tobit estimates Number of obs = 157 LR chi2(13) = 100.03 Prob > chi2 = 0.0000 Log likelihood = -353.60329 Pseudo R2 = 0.1239					

Alt. 2	Tobit Coefficients		Marginal Effects		Sig.	
	Coef.	Std. Err	dF/dX	Std. Err	t	P>t
sex	-3.35302	.9060806	-3.35302	.9060806	-3.701	0.000
age	-.0356433	.0293543	-.0356433	.0293543	-1.214	0.227
edu_prim	3.652683	1.291089	3.652683	1.291089	2.829	0.005
edu_seco	3.583043	1.567375	3.583043	1.567375	2.286	0.024
edu_post	3.853434	1.643787	3.853434	1.643787	2.344	0.020
incs_agr	-.8048257	1.011193	-.8048257	1.011193	-0.796	0.427
incs_gov	.400474	1.289544	400474	1.289544	0.311	0.757
incs_oth	-2.395335	1.773378	-2.395335	1.773378	-1.351	0.179
richer_a	1.989797	1.293691	1.989797	1.293691	1.538	0.126
poorer_a	-2.272576	1.00606	-2.272576	1.00606	-2.259	0.025
nohhmem	.2338107	.1258232	.2338107	.1258232	1.858	0.065
part_hh	.2862266	.5451377	.2862266	.5451377	0.525	0.600
dumagbe	-.7668153	.9864155	-.7668153	.9864155	-0.777	0.438
dumgef	-3.243002	1.154301	-3.243002	1.154301	-2.809	0.006
_cons	5.460053	2.467585	5.460053	2.467585	2.213	0.029
Goodness of Fit	Tobit estimates Number of obs = 157 LR chi2(14) = 95.35 Prob > chi2 = 0.0000 Log likelihood = -355.93965 Pseudo R2 = 0.1181					

---

5   Marginal effects: latent variable.

Annex 4-11: Mathematical implementation of the two alternative probit estimations

Alternative 1:

$$y_{use} = \beta_0 + \beta_1 * SEX + \beta_2 * AGE + \beta_{31} * EDU\_ILLI + \beta_{32} * EDU\_SECO + \beta_{33} * EDU\_POST + \beta_4 * LNHHINC + \beta_5 * DUMAGBE + \beta_6 * DUMGEF + u_{use}$$

Alternative 2:

$$y_{use} = \beta_0 + \beta_1 * SEX + \beta_2 * AGE + \beta_{31} * EDU\_ILLI + \beta_{32} * EDU\_SECO + \beta_{33} * EDU\_POST + \beta_{41} * POORER\_A + \beta_{42} * RICHER\_A + \beta_5 * DUMAGBE + \beta_6 * DUMGEF + u_{use}$$

Annex 4-12: Mathematical implementation of the two tobit alternatives

Alternative 1 :

$$y_{lntcexp} = \beta_0 + \beta_1 * SEX + \beta_2 * AGE + \beta_{31} * EDU\_ILLI + \beta_{32} * EDU\_SECO + \beta_{33} * EDU\_$$
$$\beta_4 * LNHHINC + \beta_5 * DUMAGBE + \beta_6 * DUMGEF + \beta_7 * NOHHMEM + \beta_{81} * INCS\_$$
$$INCS\_GOV + \beta_{83} * INCS\_OTH + PART\_HH + u_{use}$$

Alternative 2:

$$y_{lntcexp} = \beta_0 + \beta_1 * SEX + \beta_2 * AGE + \beta_{31} * EDU\_ILLI + \beta_{32} * EDU\_SECO + \beta_{33} * EDU\_$$
$$\beta_{41} * POORER\_A + \beta_{42} * RICHER\_A + \beta_5 * DUMAGBE + \beta_6 * DUMGEF \beta_7 * NOH_l$$

176

Annex 5-1: Cross tabulation of purpose of outgoing and incoming calls and community

Outgoing calls (N = 79)	Akatsi	Agbedrafor	Gefia
family, friends...	31	14	7
agric. production	1	0	0
non-agric. economic activities	11	0	0
consumer prices	1	0	0
government and administration	4	0	0
emergency and health	3	1	1
others	2	2	1
total	53	17	9

Incoming calls (N = 62)	Akatsi	Agbedrafor	Gefia
family, friends...	25	12	2
agric. production	1	1	0
non-agric. economic activities	8	0	1
consumer prices	1	0	0
government and administration	2	0	0
emergency and health	3	1	1
others	2	1	1
total	42	15	5

References

Accra Mail (2001): Mother Ghana Invites Her Children Home. March 23, 2001. Internet (04/2001): http://www.ghanaweb.com.

Agbedrafor (1999): Group interview on November, 9 with chief, elders and opinion leaders in Agbedrafor.

Akatsi District Assembly (1996): Akatsi District Assembly Medium Term Development Plan 1996-2000.

Akatsi District Assembly (1999): Group interview on April, 27 with: the District Director of Agriculture, the Presiding Member of the Akatsi District Assembly, the Agric. Extension Agent of the Ministry of Food and Agriculture, and the District Co-ordinating Director.

Akerlof, G.A. (1970): The market of lemons: Qualitative uncertainty and the market mechanism. Quarterly Journal of Economics 84, pp. 488-500.

Baily, M.N. (1986): What has happened to productivity growth? Science 234, pp. 443-451.

Bayes, A.; von Braun, J.; Akhter, R. (1999): Village Pay Phones and Poverty Reduction: Insights from a Grameen Bank Initiative in Bangladesh. ZEF-Discussion Papers on Development Policy No. 8. Bonn.

Bedi, A.S. (1999): The Role of Information and Communication Technologies in Economic Development – A Partial Survey. ZEF-Discussion Papers on Development Policy No. 7. Bonn.

Benninghaus, H. (1996): Einführung in die sozialwissenschaftliche Datenanalyse. München.

Bertolini, R. (1998): Die Rolle neuer IuK-Techniken in mittelständischen Industrieunternehmen. Mimeo. Bonn.

Bertolini R.; Anyimadu, A.; Asem; P.; Sakyi-Dawson, O. (2000): Telecommunication Services in Ghana – A Sector Overview and Case Studies from the Southern Volta Region. In: Bruene, S. (Ed.)(2000): Neue Medien und Öffentlichkeiten – Politische Kommunikation in Asien, Afrika und Latein-amerika, Band 2. Deutsches Überseeinstitut. Hamburg.

BMI-T-Group (Ed.)(1998): Communication Technology Handbook 1998. Johannesburg.

Braga, P. et al. (2000): The networking revolution – Opportunities and challenges for Developing Countries. infoDev Working Paper. Washington D.C.

Brunn, S.; Leinbach, T. (Eds.)(1991): Collapsing Space and Time: Geographic Aspects of Communication and Information. London.

Brynjolfsson, E.; Hitt, L. (1996): Paradox Lost? Firm-level Evidence on the Returns to Information Systems Spending. Management Science 42, pp. 541-558.

Bühl, A.; Zöfel, P. (1996): SPSS für Windows Version 6.1. Praxisorientierte Einführung in die moderne Datenanalyse. Bonn.

Cairncross, F. (1997): The Death of Distance. Harvard.

Capital Telecom (1999): Unpublished Company Profile. Accra.

Central Intelligence Agency (2000): World Factbook 2000. Internet (04/2001): http://www.cia.gov.

Chaterjee, S.; Price, B. (1995): Praxis der Regressionsanalyse. München.

Chibber, A.; Fischer, S. (Eds.)(1992): Economic Reform in Sub-Saharan Africa. A World Bank Symposium. Washington D.C.

Christaller, W. (1966): Central Places in Southern Germany. New Jersey.

Chu, G.C.; Srivisal , A.C.; Suphailoke, B. (1985): Rural telephone in Indonesia and Thailand. Telecommunications Policy 9, pp. 159-169.

Clark, D.; Unwin, C. (1981): Telecommunications and travel: potential impact in rural areas. Regional Studies 15, pp. 47-56.

Cleevely, D.D.; Walsham, G. (1980): Interim report on modelling the role of telecommunications within regions of Kenya. Cambridge, U.K.

Deaton, A. (1997): The Analysis of Household Surveys. Baltimore.

Dickson, K.B.; Benneh, G. (1995): A New Geography of Ghana. Harlow.

The Economist (2000): Untangling e-conomics. The Economist – A survey of the new economy. September 2000.

European Commission (Ed.)(1996): Building the European Information Society for Us All. First Reflexions of the High Level Group of Experts. Bruxelles.

Fraunhofer Gesellschaft (Ed.)(1995): Kommunikation ohne Verkehr? Neue Informationstechniken machen mobil. München.

Frempong, G.K.; Atubra W.H. (2001): Liberalisation of telecoms: the Ghanaian experience. Telecommunications Policy 25, pp. 197-210.

G8 (2000): Okinawa Charter on Global Information Society. Official Document of the Kyushu-Okinawa Summit Meeting 2000. Internet (10/2000): http://www.g8kyushu-okinawa.go.jp/e/documjents/it1.html.

G8 (2001): The Digital Opportunity Task Force. Addressing the global digital divide. Internet: (05/2001): http://www.dotforce.org.

Geertz, C. (1978): The Bazaar Economy: Information and search in peasant marketing. American Economic Review 68, pp. 28-32.

Gefia (1999): Group interview on November, 9 with chief, elders and opinion leaders in Gefia.

Ghana Statistical Service (2000): 2000 Population and Housing Census – Provisional Results. Accra.

Ghana Telecom (1998): Telephone Directory 1998. Accra.

Ghana Telecom (2000): Telephone Directory 2000. Accra.

Ghana Telecom (2000a): The Company – Basic Facts. Internet (10/2000): http://www.ghanatel.net.

Ghanaweb (1999): WESTEL Hooks Up With MOBITEL. Internet (3/1999) http://www.ghanaweb.com.

Ghanaweb (2001): Ghana Telecom expansion gets US$ 100 million loan from World Bank Arm, March 27, 2001. Internet (04/2001) http://www.ghana web.com.

Ghanaweb (2001): Draft National Information Communication Plan out in August. May 18, 2001. Internet (04/2001): http://www.ghanaweb.com.

Gillis, W.; McLellan, S. (1998): Rural Telecommunications: From Market Failure to Market Opportunity. Internet (08/2000): http://www.wutc.wa.gov/web1/misc/ruraldev.html.

Grant, I. (1998): Telecommunications policy options for Africa. In: BMI-T (1998): Communication Technologies Handbook 1998. Johannesburg.

Gujarati, D.M. (1995): Basic Econometrics. New York.

Hägerstrand, T. (1970): What about People in Regional Science? Papers of the Regional Science Association 24, pp. 7-21.

Hägerstrand, T. (1972): The Impact of Social Organization and Environment upon the Time-Use of Individuals and Households. Plan International. Special Issue, pp. 24-30.

Hamelink, C.J. (1997): New Information and Communication Technologies, Social Development and Cultural Change. UNRISD Discussion Paper No. 86. United Nations Research Institute for Social Development. Geneva.

Hanna, N.; Guy, K.; Arnold, E. (1995): The diffusion of Information Technology. Experience of industrial countries and lessons for developing countries. World Bank Discussion Paper No. 281. Washington, D.C.

Heeks, R. (1999): Information and Communication Technologies, Poverty and Development. Development Informatics Working Paper Series No. 5. Manchester.

Heuermann, A. (1999): Die Bedeutung von Telekommunikationsdiensten für wirtschaftliches Wachstum. ZEF-Discussion Papers on Development Policy No. 17. Bonn.

The Hindu (1999): Empowering villagers through information technology. Internet (05/2001): http://www.indiaserver.com/thehindu/1999/07/08/stories/08080004.htm.

Hudson, H. (1992): Telecommunications Policies for Rural Development, Policy Research Paper No. 30. Center for International Research on Communication and Information Technology, University of San Francisco. San Francisco.

IICD (1999): Draft of Communications Policy Discussion Paper. Ministry of Communications Ghana. Internet (3/1999): http://www.iicd.org.

IMF (2000): Ghana Enhanced Structural Adjustment Facility Policy Framework Paper, 1999-2001. Internet (11/2000): http://www.imf.org/external/NP/PFP/1999/Ghana.

Independent Commission for World-Wide Telecommunication Development (1984): The Missing Link: Report of the Independent Commission for World-Wide Telecommunication Development. International Telecommunications Union. Geneva.

ITU (1988): Benefits of telecommunications to the transportation sector of developing countries: A case study in the People's Republic of Yemen. Geneva.

ITU (1997): World Telecommunication Development Report (WTDR) 1996/97 – Trade in Telecommunications. Geneva.

ITU (1998): African Telecommunication Indicators (ATI) 1998. Geneva 1998.

ITU (1998a): World Telecommunication Development Report 1998 – Executive Summary. Geneva.

ITU (1998b): World Telecommunication Development Report 1998 – Universal Access. Geneva.

ITU (1999): Trends in Telecommunication Reform – Convergence and Regulation. Geneva.

ITU (2000): World Telecommunication Indicator Database 1999/2000. Geneva.

Jensen, M. (1998): The current status of the Internet and related developments in Africa. In: BMI-T (1998): Communication Technologies Handbook 1998. Johannesburg.

Jensen, M. (2000): Internet Connectivety in Africa – A status Report. Internet (11/2000): http://demiurge.wn.apc.org/africa/afstat.htm.

Johansen, R.; Vallee, J.; Spangler, K. (1979): Electronic Meetings. Reading Massachusetts.

Kilgour, M.C. (1982): The telephone in the organization of space for development. Cambridge, Massachusetts

Klingbeil, D. (1980): Zeit als Prozess und Ressource n der sozialwissenschaftichen Humangeographie. Geographische Zeitschrift 68, pp. 1-32.

Lathey, C.E. (1975): Telecommunications Substitutability for Travel: An Energy Conservation Potential.Office of Telecommunications US Dept. of Commerce, Report no. 75-58. Washington D.C.

Leff, N. (1984): Externalities, information costs, and social benefit-cost analysis for economic development: an example from telecommunications. Economic Development and Cultural Change 32, pp. 255-276.

Loveman, G.W. (1994): An assessment of the productivity impact of information technologies. In: Allen, T.J.; Morton, S. (Eds.)(1994): Information Technology and the Corporation of the 1990s. Cambridge Massachusetts

Maddala, G.S. (1983): Limited-dependent and Qualitative Variables in Econometrics. Cambridge, Massachusetts

Männistö, L.; Kelly, T.; Petrazzini, B. (1998): Internet and global information infrastructure in Africa. ITU Discussion Paper. Geneva.

Mansell, R. (1999): Information and Communication Technologies for Development: Assessing the Potential and the Risk. Telecommunications Policy 23, pp. 35-50.

Melody, W.H. (2000): Telecom development. Telecommunications Policy 24, pp. 635-638.

Meuser, M.; Nagel, U. (1991): ExpertInneninterviews – vielfach erprobt, wenig bedacht. Ein Beitrag zur qualitativen Methodendiskussion. In: Graz, D.; Kraimer, K .(Eds.)(1991): Qualitativ-empirische Sozialforschung. Konzepte, Methoden, Analysen. Opladen, pp. 441-469.

Ministry of Transport and Communications (1999): Telecommunications Policy for an accelerated development programme 1994 – 2000. Internet (3/1999): http://www.communication.gov.gh.

Ministry of Transport and Communications (2000): Telecommunications Policy for an Accelerated Development Program 1994-2000. Internet (10/2000): http://www.communication.gov.gh/telecommpolicies.htm.

Mitchell, B.M.(1978). Optimal Pricing of Local Telephone Service. American Economic Review 68, pp. 517-537.

Morrison, C.J.; Berndt, E.R. (1990): Assessing the productivity of information technology equipment in the U.S. manufacturing industries. NBER Working Paper No. 3582.

The MOSAIC Group (1998): The Global Diffusion of the Internet Project. An Initial Inductive Study. Internet (7/2000): http://www.agsd.com/gdi97/gdi97. html.

Mowery, D.C.; Oxley, J.E. (1995): Inward technology transfer and competitiveness: the role of national innovation systems. Cambridge Journal of Economics 19, pp. 67-93.

Mustafa, M.A.; Laidlow, B.; Brand, M. (1997): Telecommunication Policies for Sub-Saharan Africa. World Bank Discussion Paper No. 353. Washington.

Nelson, R.R. (Ed.)(1993): National Innovation Systems. New York.

183

Norton, S. (1992): Transactions costs, telecommunications, and the microeconomics of macroeconomic growth. Economic Development and Cultural Change 41, pp. 175-196.

One World (1997): Telecommunications development and the market: The promises and the problems. Panos Media Briefing No. 23. Internet (7/1998): http://www.oneworld.org/ panos/briefing/telecoms.htm.

Panafrican News Agency (1998): Telecom Company Slashes Subscription Fee by Half. Internet (8/1998): http://www.africanews.org/PANA/news/19980701/feat16.html.

Pohjola, M. (1998): Information Technology and Economic Development: An Introduction to the Research Issues. Helsinki.

Porter, M.E. (1990): The Competitive Advantages of Nations. New York.

Press, L. et al. (1998): An Internet Diffusion Framework. Communications of the ACM 41, pp. 21-26.

The Probe Team (1999): Public Report on Basic Education in India. Oxford.

Pye Telecommunications Ltd.: (1976): A study of the future frequency spectrum requirements for private mobile radio in the U.K. Cambridge, U.K.

Renaud, B.M. (1981): National Urbanization Policies in Developing Countries. New York.

The Republic of Ghana (1995): Ghana-Vision 2020. Presidential Report to Parliament on Co-ordinated Programme of Economic and Social Development. Accra.

Ritter, W. (1998): Allgemeine Wirtschaftsgeographie. München.

Roach, S.S. (1991): Under siege – the restructuring imperative. Harvard Business Review, pp. 82-92.

Roche, M.; Blaine, M.J. (1996)(Eds.): Information Technology, Development and Policy. Avebury, U.K.

Rogers, E. M. (1995): Diffusion of Innovations. New York.

SAPRIN (2000): Draft SAPRI Methodological Framework. Internet (11/2000): http://www.igc.org/dgap/saprin/meth3a.html.

Satellite Communication Services (1981): Teleconferencing Newsletter 1, No. 1.

Saunders, R. J., Warford, J. J., Wellenius, B. (1994): Telecommunications and Economic Development. Washington D.C.

Scherer, P. B. (1994): Telecommunications Reform in Developing Countries. Importance and Strategy in the Context of Structural Change: In: Wellenius, B.; Stern, P.A. (Eds.)(1994): Implementing Reforms in the Telecommunications Sector. World Bank Regional and Sectoral Studies. Washington D.C.

Schmidt-Kallert, E. (1994): Ghana. Gotha.

Schnell, R.; Hill, P.; Esser, E. (1992): Methoden der empirischen Sozialforschung. München.

Seibel, S.; Müller-Falcke, D.; Bertolini, R. (1999): Informations- und Kommuni-kationstech-nologien in Entwicklungsländern. ZEF-Discussion Papers on Development Policy No. 4. Bonn.

Shannon, G. W. (1997): Telemedicine: Restructuring Rural Medical Care in Space and Time. In: Bashshur, R. L. et al. (Eds.)(1997): Telemedicine: Theory and Practice. Springfield.

Stiglitz, J.E. (1988): Economic Organization, Information, and Development. In: Chenery, H.; Srinivasan T.N. (Eds.)(1988): Handbook of Development Economics 1, pp. 93-160.

Talero, E.; Gaudette, P. (1995): Harnessing Information for Development. A Proposal for a World Bank Group Strategy. Washington D.C.

Tarjanne, P. (1999): Preparing for the next revolution in telecommunications: implementing the WTO agreement. Telecommunications Policy 23, p. 51-63.

Telecommons Development Group (2000): Rural Access to Information and Communication Technologies: The Challenge of Africa. African Connection Secretariat 2000.

Tetsch, F. (1985): Zur regionalen Bedeutung der neuen Techniken zur Individual-kommunikation (Telematik). Raumforschung und Raumordnung 6/85, pp. 270-278.

Thiem, M. (1998): Zur Situation indianischer Desplazados in der Stadt Guatemala. Mimeo. Cologne.

Torero, M. (2000): The Access and Welfare Impacts of Telecommunications Technology in Peru. ZEF-Discussion Papers on Development Policy No. 27. Bonn.

Tyler, M. (1978): Implications for Transport. In: Smith, R.C. (Ed.)(1978): Impacts of Telecommunications on Planning and Transport. Research Report No. 24. London.

UNDP (1998): Human Development Report 1998. New York.

Varian, H.T. (1987): Intermediate microeconomics. New York.

Varian, H.T. (1995): Grundzüge der Mikroökonomik. München.

Webber, M. (1980): A telecommunications strategy for new cities of the 21st century. Working Paper No. 330. Berkeley.

Weinberger, K. (2000): Women's Participation: an Economic Analysis in Rural Chad and Pakistan. Frankfurt.

Wellenius et al. (2000): Investment and growth of the information infrastructure: summary results of a global survey. Telecommunications Policy 24, pp. 639-643.

Williamson, O.E. (1979): Transaction-Cost Economics: The Conveyance of Contractual Relations. Journal of Law and Economics 22, pp. 426-448.

Wolfensohn, J. D. (1999): A Proposal for a Comprehensive Development Framework. Unpublished Report to the Board, Management and Staff of the World Bank Group. A Discussion Draft. Washington D.C.

The World Bank (1998): World Development Report 1998/99. Knowledge for Development. Washington D.C.

Development Economics and Policy

Edited by Franz Heidhues and Joachim von Braun

Band 1 Andrea Fadani: Agricultural Price Policy and Export and Food Production in Cameroon. A Farming Systems Analysis of Pricing Policies. The Case of Coffee-Based Farming Systems. 1999.

Band 2 Heike Michelsen: Auswirkungen der Währungsunion auf den Strukturanpassungsprozeß der Länder der afrikanischen Franc-Zone. 1995.

Band 3 Stephan Bea: Direktinvestitionen in Entwicklungsländern. Auswirkungen von Stabilisierungsmaßnahmen und Strukturreformen in Mexiko. 1995.

Band 4 Franz Heidhues / François Kamajou: Agricultural Policy Analysis – Proceedings of an International Seminar, held at the University of Dschang, Cameroon on May 26 and 27 1994, funded by the European Union under the Science and Technology Program (STD). 1996.

Band 5 Elke M. Förster: Protection or Liberalization? A Policy Analysis of the Korean Beef Sector. 1996.

Band 6 Gertrud Schrieder: The Role of Rural Finance for Food Security of the Poor in Cameroon. 1996.

Band 7 Nestor R. Ahoyo Adjovi: Economie des Systèmes de Production intégrant la Culture de Riz au Sud du Bénin: Potentialités, Contraintes et Perspectives. 1996.

Band 8 Jenny Müller: Income Distribution in the Agricultural Sector of Thailand. Empirical Analysis and Policy Options. 1996.

Band 9 Michael Brüntrup: Agricultural Price Policy and its Impact on Production, Income, Employment and the Adoption of Innovations. A Farming Systems Based Analysis of Cotton Policy in Northern Benin. 1997.

Band 10 Justin Bomda: Déterminants de l'Epargne et du Crédit, et leurs Implications pour le Développement du Système Financier Rural au Cameroun. 1998.

Band 11 John M. Msuya: Nutrition Improvement Projects in Tanzania: Implementation, Determinants of Performance, and Policy Implications. 1999.

Band 12 Andreas Neef: Auswirkungen von Bodenrechtswandel auf Ressourcennutzung und wirtschaftliches Verhalten von Kleinbauern in Niger und Benin. 1999.

Band 13 Susanna Wolf (ed.): The Future of EU-ACP Relations. 1999.

Band 14 Franz Heidhues / Gertrud Schrieder (eds.): Romania – Rural Finance in Transition Economies. 2000.

Band 15 Katinka Weinberger: Women's Participation. An Economic Analysis in Rural Chad and Pakistan. 2000.

Band 16 Christof Batzlen: Migration and Economic Development. Remittances and Investments in South Asia: A Case Study of Pakistan. 2000.

Band 17 Matin Qaim: Potential Impacts of Crop Biotechnology in Developing Countries. 2000.

Band 18 Jean Senahoun: Programmes d'ajustement structurel, sécurité alimentaire et durabilité agricole. Une approche d'analyse intégrée, appliquée au Bénin. 2001.

Band 19 Torsten Feldbrügge: Economics of Emergency Relief Management in Developing Countries. With Case Studies on Food Relief in Angola and Mozambique. 2001.

Band 20 Claudia Ringler: Optimal Allocation and Use of Water Resources in the Mekong River Basin: Multi-Country and Intersectoral Analyses. 2001.

FSC
www.fsc.org
MIX
Papier | Fördert
gute Waldnutzung
FSC® C083411

Zeitfracht Medien GmbH
Ferdinand-Jühlke-Straße 7
99095 Erfurt, Deutschland
produktsicherheit@kolibri360.de

Druck:
CPI Druckdienstleistungen GmbH
im Auftrag der
Zeitfracht Medien GmbH
Ein Unternehmen der Zeitfracht - Gruppe
Ferdinand-Jühlke-Str. 7
99095 Erfurt